School Food Politics

A.C. (Tina) Besley, Michael A. Peters,
Cameron McCarthy, Fazal Rizvi
General Editors

Vol. 6

The Global Studies in Education series is part of the Peter Lang Education list.
Every volume is peer reviewed and meets
the highest quality standards for content and production.

PETER LANG
New York • Washington, D.C./Baltimore • Bern
Frankfurt • Berlin • Brussels • Vienna • Oxford

School Food Politics

THE COMPLEX ECOLOGY OF HUNGER AND FEEDING IN SCHOOLS AROUND THE WORLD

EDITED BY Sarah A. Robert AND Marcus B. Weaver-Hightower

PETER LANG
New York • Washington, D.C./Baltimore • Bern
Frankfurt • Berlin • Brussels • Vienna • Oxford

Library of Congress Cataloging-in-Publication Data

School food politics: the complex ecology of hunger and feeding
in schools around the world /
[edited by] Sarah A. Robert, Marcus B. Weaver-Hightower.
p. cm. — (Global studies in education; v. 6)
Includes bibliographical references and index.
1. School children—Food. 2. Food—Political aspects. I. Robert, Sarah A.
II. Weaver-Hightower, Marcus B. III. Title. IV. Series.
LB3475.S36 371.7'16—dc22 2011007640
ISBN 978-1-4331-1308-6 (hardcover)
ISBN 978-1-4331-1307-9 (paperback)
ISSN 2153-330X

Bibliographic information published by **Die Deutsche Nationalbibliothek**.
Die Deutsche Nationalbibliothek lists this publication in the "Deutsche
Nationalbibliografie"; detailed bibliographic data is available
on the Internet at http://dnb.d-nb.de/.

Cover photos: Bottom right © Doug Davis.
All others © Marcus B. Weaver-Hightower.

The paper in this book meets the guidelines for permanence and durability
of the Committee on Production Guidelines for Book Longevity
of the Council of Library Resources.

29 Broadway, 18th floor, New York, NY 10006
www.peterlang.com

Printed in the United States of America

For Julia and Lilah, Sarah's "good" eaters

For Harrison, Marcus' son, who has to eat the stuff

• CONTENTS •

Acknowledgments ix

Foreword xi
Chef Ann Cooper

Introduction: School Food Politics 1
Marcus B. Weaver-Hightower & Sarah A. Robert

SECTION ONE
From Pap to Sloppy Joes to Nada: Inside International School Food Politics

1 Reframing the Politics of Urban Feeding in U.S. Public Schools: Parents, Programs, Activists, and the State 25
Jen Sandler

2 Fixing Up Lunch Ladies, Dinner Ladies, and Canteen Managers: Cases of School Food Reform in England, the United States, and Australia 46
Marcus B. Weaver-Hightower

3 Cultivating Schools for Rural Development: Labor, Learning, and the Challenge of Food Sovereignty in Tanzania 71
Kristin D. Phillips & Daniel Roberts

4 Defining the "Problem" with School Food Policy in Argentina 94
Sarah A. Robert & Irina Kovalskys

5 Free for All, Organic School Lunch Programs in South Korea 120
Mi Ok Kang

SECTION TWO
Reforming School Food: Parents, Activists, Teachers, and Youth

6 School Food, Public Policy, and Strategies for Change 143
Marion Nestle

7 Food Prep 101: Low-Income Teens of Color Cooking Food and Analyzing Media 147
Catherine Lalonde

8 Going Local: Burlington, Vermont's Farm-to-School Program 162
Doug Davis, Dana Hudson, & Members of the Burlington School Food Project

9 What's That Non-Human Doing on Your Lunch Tray? Disciplinary Spaces, School Cafeterias, and Possibilities of Resistance 183
Abraham DeLeon

10 Coda: Healthier Horizons 201
Sarah A. Robert & Marcus B. Weaver-Hightower

Appendix: School Food Resources: From Curriculum to Policy to Recipes 209

Contributors 213

Index 213

• ACKNOWLEDGMENTS •

This collection is a result of a serendipitous collision of work on school food. Sarah and Marcus were friends from graduate school who got a panel together on the usual subject of their work—gender—for the annual conference of the Comparative and International Education Society in 2008. As it happened, Sarah, Kristin Phillips, and Jen Sandler (the latter two also contributors to this volume) had organized a panel on food issues for the same conference. Marcus had recently gotten a small grant to begin a study on school food himself, and so he sat in the audience for this food panel and was enthralled by the papers that Sarah, Kristin, and Jen gave. Neither Sarah nor Marcus knew the other was working on food politics, but it was gratifying to find solidarity in these common interests. Ensuing conversations (and a lot of work) gave rise to this book.

Though it might sound trite, this book could not have happened without much help. The assistance and support of hundreds of people (a vast ecology!) have shaped what you see before you. Both of us would like to thank those who supported both our individual contributions and thc compilation of the overall collection.

Specifically, Sarah would like to thank Heather McEntarfer, University at Buffalo (UB) graduate assistant and scholar par excellence, who has pulled this compilation together from literature searches to editing text and documents. Marie Kleiderlein, another UB graduate assistant, provided additional assistance and English expertise in the final stages of the project. Christopher Hollister, Education Librarian at UB's Lockwood Library, is owed so much for his savvy with databases and overflowing kindness even on tight schedules. She is indebted to Michael Apple, Lois Weis, and Greg Dimitriadis, who offered sage advice on this project at its earliest stages. She thanks Julia for her wise fourth grader critiques of school food politics and regular inquiries to find out if Mommy was finished with "this" book yet. She is grateful for Lilah's company at the computer, enthusiastically sitting down next to Mommy to do her pre-kindergarten "homework." Last, she is forever grateful to Nicolás Penchaszadeh, her partner, ever patient and firm in his demand for "the lead," even for academic stories, while leading our household.

Marcus thanks, first and most importantly, those who participated in the research as interviewees and "observees." Bekisizwe Ndimande, Becky Francis, Nicola Tilt, Jori Thordarson, and the staffs at the National Library of Australia, the U.S. Library of Congress, the British Library, and the University of North Dakota's Chester Fritz Library have helped in numerous ways to find him participants and information. He also thanks his students at the University of North Dakota for their help, for listening to his in-process ideas and adding many of their own. This research was supported by a New Faculty Scholar Award from the University of North Dakota. It was also supported by the generous support and advice of his departmental colleagues, including Kathy Gershman, Richard Kahn, Dick Landry, and Steve LeMire. He thanks his son, Harrison, for patiently waiting for Daddy to type on his "'puter" before he could play superheroes. My angel babies were inspiration, too. Most especially, Marcus thanks his wife, Rebecca (who "looks like a movie star" and is a true scholar), for her patience and for the extra work of covering for some late nights of writing and then reading it when it was done.

Some portions of this book were also presented at the annual meeting of the American Educational Research Association in Denver in 2010.

Portions of the Introduction appear in different form in Marcus Weaver-Hightower's "Why Educational Researchers Should Take School Food Seriously," in *Educational Researcher* (volume 40, number 1, pp. 15–21; doi: 10.3102/0013189X10397043).

"School Food, Public Policy, and Strategies for Change" by Marion Nestle was originally published by the Center for Ecoliteracy. © Copyright 2004 Center for Ecoliteracy. Reprinted with permission. All rights reserved. For more information, visit www.ecoliteracy.org.

• FOREWORD •

I Am Such an Unlikely Lunch Lady!

Chef Ann Cooper,
the "Renegade Lunch Lady"

I never saw myself becoming a lunch lady, and I still often ponder how I found myself here in Boulder, Colorado, feeding thousands of hungry children daily as a part of a government-subsidized meal program.

But I'm here because this is what I was meant to do; it's a passion I have followed since awakening to the historical inadequacy that is our National School Lunch Program in the United States. While writing my second book, *Bitter Harvest* (2000), which speaks to the relationship between food, politics and health, I came to understand that food often makes us and our kids sick. Through my research on the subject, I uncovered the deep connection between agribusiness and our government—a connection that often results in deleterious effects on our health, our kids' health, and the health of the planet.

The corruption in this connection between government and agribusiness affects all of us, but it has particularly created unexpected negative consequences for the National School Lunch Program. Bad foods turn up on students' trays as we serve the same four meals over and over again: pizza, chicken nuggets (or some form of mashed chicken and chicken by-product), hamburgers, hot dogs (or corn dogs!), pizza, chicken nuggets, hamburgers, hot dogs, pizza, chicken nuggets, hamburgers, hot dogs, pizza, chicken nuggets, hamburgers, hot dogs ... all with a side of fries or tater tots. Bland, fried, and nutrient-deficient food.

I personally could not stand by idly as children stuffed their faces with food we would not even feed to our country's prisoners. That is why I became a lunch lady.

When I entered the realm of school food, I found kids of all sizes—malnourished children who are too thin *and* malnourished children who are too big. I had to ask myself How did things get this way?

Why don't we trust our children to eat and enjoy healthy foods? Why did we not seem to notice the connection between what we were feeding students and their behavior in class? And when our kids started getting BIG and developing problems that come with having lots of big kids—diabetes, asthma, and behavior disorders such as Attention Deficit and Hyperactivity Disorder—why didn't we look at changes in the sources of our food as a possible cause?

How did we not listen as the U.S. Centers for Disease Control and Prevention (CDC, 2007) stated that, of the children born in the year 2000, one out of every three Caucasians and one out of every two African Americans and Hispanics will contract diabetes in their lifetimes? Further, those same children may be part of the first generation in U.S. history to die at a younger age than their parents (Olshansky et al., 2005), all because of what we feed them. We as adults—educators and caregivers—are the responsible party in the denigration of our children's health. We are feeding our children to death, and we truly must stop!

Changing the way children eat should be the social action cause of the century. Reversing the rut we are stuck in—spending too little money on poor quality foods, dismissing food and healthy eating education, allowing our kids' taste buds to get accustomed to high-fat, high-sugar, high-salt meals—is a difficult task. But it is possible, and I have seen it happen. I have witnessed "picky" eaters take full advantage of the salad bars now in place at all 48 schools of the Boulder Valley School District. I noticed the questions of "Where is my chocolate milk?" dissipate as students began to enjoy the fresh, local and organic milk now served. And I watched as even the toughest converts started getting used to and even enjoying healthy food.

This collection, *School Food Politics: The Complex Ecologies of Hunger and Feeding in Schools Around the World*, has arrived at a much-needed time in our country's history. With interest in the state of food policy consistently growing in the United States and across the world, the topic of *school food*, particularly, has earned its spot in lesson plans across an array of college courses. By educating oneself on the topic of school food, you become an advocate. If you care about the health of our children—*your* children—school meals become an increasingly interesting and pressing subject. Studying school meals across the United States and the world, as *School Food Politics* does, will enable action to take place and change to happen.

The array of subjects covered in *School Food Politics* allows the reader to get an in-depth look and feel for the food stories, policies, and people that surround this topic. By describing school food programs across the world, this book enables us to not only consider how other countries (both wealthy and poor, democratic and not) deal with feeding schoolchildren, but also how we can apply positive aspects of other systems to our own here in the United States. This subject stretches beyond politics and explores everything from family dynamics to science and nutrition, from technological advances to simply creating delicious food.

I know firsthand the difficulty of making school lunch better. When I began helping schools to make these changes, a reporter nicknamed me the "Renegade Lunch Lady." While it's a cute and catchy title, I dream of the day when what I'm doing will not be considered so "renegade." I dream of the day when all of America's schoolchildren walk into their cafeterias to find homemade, delicious, flavorful, and just plain *good* food. I dream of the day when lunch ladies can feel confident in the healthy food they place in front of children. And I dream of a day when we begin to see improvement in grades, behavior, and self-image—along with a myriad of other positive benefits—that come from feeding our students better food!

Discussing food policy leads to changes in food policy, and *School Food Politics* will spark discussions that will turn into changes for schoolchildren across America and the globe.

References

Centers for Disease Control and Prevention. (2007). *National Diabetes Surveillance System. Incidence of Diabetes: Crude and Age-Adjusted Incidence of Diagnosed Diabetes per 1000 Population Aged 18–79 Years, United States, 1997–2004.* Retrieved April 17, 2007, from http://www.cdc.gov/diabetes/statistics/incidence/fig2.htm.

Cooper, A., & Holmes, L. M. (2000). *Bitter harvest: A chef's perspective on the hidden dangers in the foods we eat and what you can do about it.* New York: Routledge.

Olshansky, S. J., Passaro, D. J., Hershow, R. C., Layden, J., Carnes, B. A., Brody, J. et al. (2005). A potential decline in life expectancy in the United States in the 21st century. *New England Journal of Medicine, 352*(11), 1138.

• INTRODUCTION •

School Food Politics

Marcus B. Weaver-Hightower
Sarah A. Robert

Obese children, tainted burgers, vending machines filled with "junk" food, and the prevalence of pre-packaged, high-fat "carnival fare" school lunches have been a major subject of debate in highly developed countries in recent years. In developing and economically constrained countries, on the other hand, distributions of fortified biscuits, micronutrient supplements, in-school rations, take-home rations of cooking oil or grain, and deworming have been of equal concern. Still, in both developed and developing contexts, the two concerns co-exist: fighting obesity and hunger simultaneously. Opinions and reform movements have been diverse, and the debates have been heated. Anti-hunger advocates have fought those who believe "handouts" breed dependence and laziness. Advocates for more organic, healthy fare have been lambasted by conservative pundits as "food police" with elitist aims. Food manufacturing corporations have fought to preserve their public image while health promoters, animal activists, anti-obesity campaigners and small-scale farmers have sought to expose those corporations' supposed excesses. We can thus see that how and what food students are fed, when and where, and—too often—*whether* students are fed at all are cultural, social, religious, economic, and ideological issues with dramatic impact on schools' manifest mission: educating children. In short, school food is political.

"School food politics," the foundation of our title and the collection's focus, is a striking, almost discordant phrase. It juxtaposes something many educationalists rarely think of—school food—with something most live every day and take quite seriously—politics. Until now, most educators and researchers of education have overlooked school food and its politics, yet increasingly the political conflicts and settlements around school food are hard to ignore, and the consequences (both real and feared) are hard to accept. This book seeks to unpack the important juxtaposition of school food and politics, giving examples from all over the world of people and groups who confront the challenges and wage the debates provoked by providing food at school.

The essays in *School Food Politics: The Complex Ecology of Hunger and Feeding in Schools Around the World* explore the intersections of food provision in schools and politics on all six of the inhabited continents of the world. Including electoral fights over universally free school meals in Korea, nutritional reforms to school dinners in England and canteens in Australia, teachers' and doctors' work on school feeding in Argentina, and more, the chapters in the first section provide key illustrations of the many contexts that have witnessed the deep, intense struggles over defining which children will eat, why, what and how they are served, and who will pay for and prepare it. The second section, then, focuses on reformers from their own perspectives. From the farm-to-school program in Burlington, Vermont, to efforts to apply principals of critical pedagogy in both cooking programs for urban teens and in animal rights curriculum, these chapters shift our attention to possibilities and hope for a different future for school food, one that is friendlier to students, "lunch ladies," society, other creatures, and the planet.

Before providing an overview of the individual chapters, let us first outline the perspectives that guide the entire collection. Each chapter coheres with the others because all are built around the notions that school food is a part of a larger food politics and that understanding this politics and solving its challenges requires thinking about school food as a complexly interrelated ecology.

Food Politics

Just what do we mean when we refer to "food politics" (ignoring for now the "school" part of the titular phrase)? We rely to a large extent on Paarlberg's (2010) definition:

> The struggle over how the losses and gains from state action are allocated in the food and farming sector is what we shall call *food politics*. The distinctive feature is not simply social contestation about food but the potential engagement of state authority. If you and I disagree over the wisdom of eating junk food, that is not food politics. If you and your allies organize and take political action to impose (or block) new government regulations on junk food...that is food politics. (p. 2; italics original)

Pointing to the involvement of the state (at whatever level) is a key understanding about *school* food politics, as well. It's not just "contestation" that provokes the concern and ire over school food, though there certainly is much contestation. Rather, school food politics is fiercely fought over because it involves the state and the real and symbolic violence (Bourdieu & Passeron, 1990) that the state can wield.

Think of it this way: no one gets very upset *that* children eat at school (excepting, of course, rules against gum or against having soft drinks around carpets and computers). Instead, it is *where* food comes from, *when* and *what* children eat, *how much* it costs, and *why* things are the way they are that creates problems (see Sandler's chapter in this volume). These are all questions of politics because a state institution, the school, is ultimately making these determinations.

While the state is indeed intimately involved in nearly every aspect of food policy, it is insufficient to focus only on the nation-state's mediation of food choices and provision. As Lien and Nerlich argue in the introduction to their volume *The Politics of Food* (2004), the complexities of the food system require "that our notion of the politics of food is expanded to the fields and arenas not traditionally thought of as 'political'" (p. 2), that is, beyond what political institutions do. The contributors to this volume take up this expansion of fields and arenas.

First, to expand our vision of food politics, we see the politics of food as a *transnational* issue, influenced by organizations "above" the authority of any one nation-state. This includes both organizations that are nominally political, like the United Nations' Food and Agriculture Organization (FAO), and those that are privately operated, like major international corporations (Monsanto, ADM, Cargill, McDonald's, Coca-Cola, etc.). Such extra-governmental organizations are not beholden to the normal democratic processes (what we would normally think of as "politics") that nation-states or even local governments are, and the invisible hand of the market is no substitute for true democracy. If Monsanto succeeds in eliminating all but its own genetically modified seeds from the market (Patel, 2007), for example, it hardly matters what any one country's government has to say, much less what a neighborhood school food reform group wants. Yet this is still food politics.

Second, we also see the politics of food as a *local, interactional* issue, operating "below" the level of the state. If we conceive of the resources of power as multiple—as economic, symbolic, and cultural capital (Bourdieu, 1977) alongside resources of literal violence—then it is easier to see that the state does not have a monopoly on "politics." In other words, individuals acting outside of state control are also involved in "the distribution of social goods"—Gee's (2005) definition of politics—because they can wield various forms of capital and change how things are done regardless of what the state says. Large-scale philanthropy by Bill Gates and others provides a good example; a grant from the Gates Foundation (e.g., Shear et al., 2008) can in some cases shape state policy in far more profound ways than local activism, giving Bill Gates's priorities an advantage over public priorities because of his economic capital. Even at a local level, though, in any neighborhood or community, particular actors have more connections, more money, and hegemonic positions (based on local politics of race, class, gender, sexuality, religion, and much more), and—often regardless of the state—can influence what kids learn or eat at school.

Consider four simple anecdotes:

1. When I (Marcus) was in elementary school in the South Carolina suburbs of the 1970s, it was already uncool to eat the school lunch. Instead, if your mom really loved you—according to the six- or seven-year-old kid logic holding sway—she would pack you a lunch. And not just any packed-lunch peanut butter and jelly sandwich would do; the more well-off kids with mothers who were, as the advertising said, "choosy" (and who, again, loved them) ate only JIF brand peanut butter on white bread. My mom, who always worked full time when I was a child, on the rare occasions that she would make me a sack lunch, always used generic brand on wheat bread instead. It was sad to realize that my mom didn't "love" me and that we were too poor to buy JIF.
2. While observing a history class in a working poor neighborhood in Metropolitan Buenos Aires, the maintenance staff entered to distribute sandwiches and juice boxes to all students. Some ate immediately. Some hid the sandwich away in their bag or desk. Still others refused it, and the staff member just left it on the desk in front of them. "See that? Did you see that?" the teacher asked me (Sarah). "They are so proud they will not even accept free food even though they need to eat."
3. Kids at Mamelodi Primary School in a township just outside of Pretoria, South Africa, have a universal school meal provided under the National School Nutrition Program (NSNP), the "universal" meaning that everyone, regardless of their income, can eat. They don't have to have tickets or provide any income proof; the children just go and eat. The food is extremely basic, usually some variant of a porridge made from corn. The food is made in a brick shelter on what looks like huge camp stoves in five gallon steel pots by women who earn about 700 rand per month (about US$91) for doing this heavy labor five days a week. The school's uniformed children have their brightly colored, round plastic plates piled with the food, they eat out in the courtyard with no chairs in sight, and then they wash their own plates in a tub of increasingly gray water and disappearing suds. Leftovers are given to the poorest children to take home for their families; this, one deputy principal told me (Marcus), keeps some of these kids coming to school rather than hanging out in the streets. Though most did, not all the kids were eating the feeding scheme

food. Others brought lunch boxes from home instead, which they told me often include sandwiches or hot dogs. Though the deputy principal told me that there is no stigma for the children who eat the free lunches, when we went to one of the seventh grade classes and asked about lunch boxes, I got the strong impression that it was far better—more prestigious—to bring a lunch from home. Those who intimated this, though, looked sheepish and smiled slyly, as if embarrassed. This chagrin was not surprising: with such poverty all around, it would seem cruel of them to suggest in front of their peers that eating the school food was a bad thing.

4. As a high school student in the 1980s who was eligible for reduced price meals, I (Sarah) recall the lunch ladies cautiously hiding that I only paid a dime. They never asked me for the set fee and they quickly popped open and closed the register to clear the ringed total. They were seemingly more aware than I of the potential stigma of paying less. My family had moved from a poor urban district where the vast majority of students paid low or no fee to a wealthier suburb where a small percentage of students did. I just wanted the food; I had to be smart about eating when I could for as cheaply as possible.

These vignettes illustrate just some of the local complexities of school lunch. In all of them we see school food politics played out within the local, sociopolitical lives of children, where they suture together issues of gender (mothers' roles in preparing food or "lunch ladies" providing it), class (who can afford sack lunches or particular brand names and who could just avoid eating), and wanting acceptance through food (see also Ludvigsen & Scott, 2009), sometimes mediated through the corporate advertising that supported these notions or through the norms of the communities students are situated within. In the second and third vignettes, we find contrasts with "Western" ideas of school food, including of course the menu and lack of choice. Yet we also see familiar issues about labor and gender (who does the work of feeding kids and how much are they are paid); national policies (South Africa has few "charity" policies outside of the NSNP and they don't do "means testing" like so many countries); and particularly issues of poverty and hunger (a battle waged every day among their friends and neighbors) amidst emerging class politics as more post-apartheid black South Africans begin reaching the middle class or, in the case of Argentina, where more and more families fall below the middle class and working class in the wake of fiscal crisis. All of these issues are part and parcel of why school food is so political, why it is a battleground in the distribution of social goods and prestige by competing members of society. In some cases this directly involves the state, but in many ways it does not.

This is what we mean when we refer to food politics, then: the combined working of the state alongside the local and transnational influences on a particular context for the growing, manufacturing, distribution, instruction about, and consumption of food. When we apply this definition to schools, particularly, we find that school food is a political topic *par excellence*. Because it involves children—and *all* children eat, no matter their "track," disability, gender, age, or other ways we sort them—the stakes are high for school feeding. Moreover, food is a very personal issue. On a physical level, food impacts children's bodies. On a cultural level, it impacts children's minds and spiritual lives. If one takes to heart Brillat-Savarin's oft-quoted adage "Tell me what you eat and I will tell you what you are," it is no wonder parents, students, educators, and communities fight so hard over whether and what their children are fed; food is a direct link to what their children are and will be.

The Policy Ecologies of School Food

School food is complicated and, as we have argued, it is deeply political on every level. Still, to most people, school feeding seems simple enough: you hire some workers, buy some food, cook it, and serve it to kids. This happens most days of the year in hundreds of thousands of schools around the world. When you dig deeper, though, you find that each stage of the process, from growing the food to spooning it onto a child's plate, is part of a vast, complex, political *ecology* involving hundreds of millions of individual actors worldwide who have varied and often competing interests and desires.

To better understand such complex interconnections that make school food so political and thus so hard to reform, the contributors to *School Food Politics* have all used—with varying degrees of explicitness—a framework of policy as an ecology (Weaver-Hightower, 2008). Based on a metaphor of ecosystems in nature, a "policy ecology"

> centers on a particular policy or related group of policies, both as texts and as discourses, situated within the environment of their creation and implementation. In other words, a policy ecology consists of the policy itself along with all of the texts, histories, people, places, groups, traditions, economic and political conditions, institutions, and relationships that affect it or that it affects. Every contextual factor and person contributing to or influenced by a policy in any capacity, both before and after its creation and implementation, is part of a complex ecology. (p. 155)

To analyze a policy ecology in this way requires, first, treating as policy both those texts that are explicitly called policy as well as those texts that act in the capacity of policy, even when they officially are "just" reports, traditions, or even verbal edicts. Second, the analyst then explores the *actors, relationships, environments and structures,* and *processes* of which an ecology is comprised.

The *actors* are all those who fill the varied roles necessary to maintain the ecology, and many actors perform multiple roles (a cafeteria worker can also be a parent and a reform activist, for example). Of course, as noted earlier, certain actors have more capital—more power—than others to effect change, and identifying and analyzing these actors is a key task of a policy ecology analysis.

The *relationships* in an ecology fall into four basic types: competition, cooperation, predation, and symbiosis. In other words, actors can, respectively, work against one another, with one another, for the destruction of the other, or off one another. Understanding these relationships helps define the tenor of the channels of power in an ecology and can show the social resources particular groups have or the barriers they are up against in their fights for reform.

The *environments and structures* of an ecology similarly define the resources and barriers for actors and their relationships. These include the ecology's (a) *boundaries*, or limits within which the actors work and the policies apply; (b) *extant conditions,* or the natural and human socio-cultural environment at the time of analysis; (c) *pressures* toward change; (d) *inputs* like money, time, materials, technology, and so on; (e) amounts and rates of *consumption* of those resources; (f) *niches and roles* that actors can take up; (g) *adaptive decentralization,* or the lack of centralized control—be it the state or otherwise—over the ecology; and (h) *agency*, the ability to act within and potentially change the ecology. All of these environments and structures shape what already exists—the status quo—and what is possible for reforms.

Finally, the *processes* in an ecology are the dynamics that occur as time unfolds and conditions change. These processes include the *emergence* of new ecologies, the *entropy* or breakdown of extant ecologies, *adaptation* to altered conditions, *conversion* to a dramatically different form, the *fragmentation* or splitting of an ecology into multiple ecologies, the *succession* or replacement of an ecology by a new one, *conservation* of resources, *anticipation* of changes, and the creation of *redundancy* in roles. Because policy ecologies are never stable or fixed, but instead are always in flux, analysts must track such processes to understand why some actors succeed and others do less well.

To illustrate how we might understand school food with this ecological metaphor, consider Figure 1. This represents one possible ecology, modeled on a school in the U.S. state of North Dakota. (Each ecology could be similarly illustrated, and some overlaps would be present across all such illustrations given the transnational nature of food policy.) On the left side of the bottom half of the figure, there is a box representing the local food ecology. The "policy" in question here is the lunch being served—the tray in the middle of the smaller dotted-line box—in one school or district (which in North Dakota can be one and the same). A number of actors have to engage in cooperative relationships to make the local system work, including the administrators, nutrition directors, and cafeteria staff that run the food program. Parents, teachers, and students participate through consuming the meals, paying for them, and, in some cases, giving direction for what they would like the programs to serve.

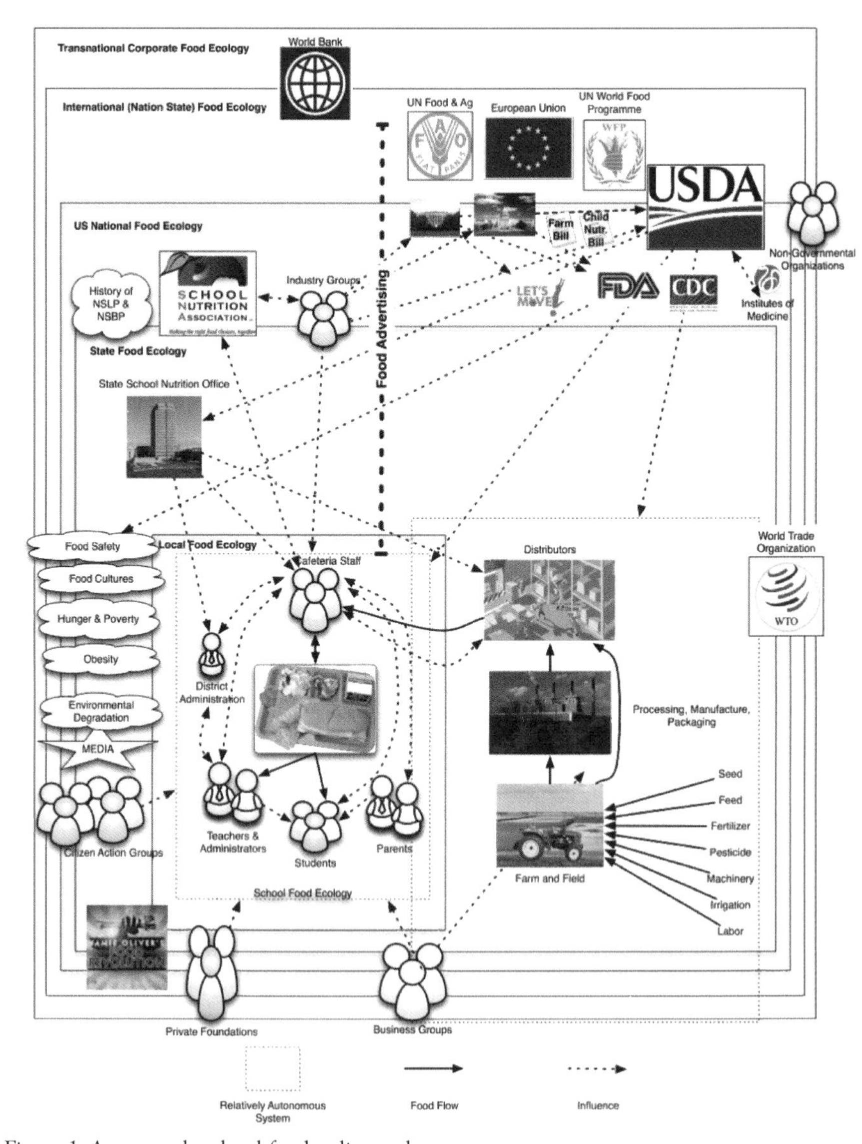

Figure 1. An example school food policy ecology.

Outside of this ultra-local context are yet more actors and relationships—lines of influence—that define the boundaries of possibility for the local system. Farmers, manufacturers, and distributors (in the box to the immediate right) supply the food for the local ecology, and what they offer and the practices they use circumscribe possibilities within the local ecology. If, for instance, the local food broker doesn't offer veggie burgers (or doesn't offer them at a reasonable price; see Weaver-Hightower's chapter, this volume), then the vegetarians in a school will have more limited choices. The local ecology is also beholden to North Dakota's nutrition bureaucracy (just above the local box in the diagram), for, in the United States, the states are in charge of administering the National School Lunch Program (NSLP), a result of race politics and states' rights politics during the creation of the program (Levine, 2008). There are also state-level and national groups that vie for policy attention, funding, and status, including the School Nutrition Association (SNA) and its state branches, food industry groups (e.g., Nestle, 2007; see also her chapter, this volume), citizen action groups, private foundations, and more.

At the federal level in the United States (in the upper right half of the figure), numerous institutions within and affiliated with the government have dominion over school food programs and shape what cafeterias can and cannot do. These include the U.S. Department of Agriculture, which runs the NSLP and coordinates a program of agricultural surplus commodities; the Food and Drug Administration, which oversees food labeling and food safety; the Centers for Disease Control and Prevention, which does research and advising on food and wellness issues; and the Institutes of Medicine, which is a major advisor to Congress on what nutritional requirements should be made in the NSLP. Each of these institutions works inside multiple legislative frameworks, including most directly the Farm Bill (see Imhoff, 2007), the Child Nutrition Act, and the Richard Russell National School Lunch Act, alongside other indirectly related legislation, including labor, civil rights, work safety, food safety, health, education, and transportation laws. (For a comprehensive understanding of the U.S. context for school food, see Poppendieck, 2010.)

The transnational and international levels of the ecology (at the top and the far sides of the figure) have perhaps fewer direct influences on the local level in the United States than in other countries. Particularly in less developed countries that receive food aid, non-governmental organizations and arms of the United Nations can be highly involved. World Bank and International Monetary Fund loans can put tight constraints on or even help create meal systems in countries, often through structural adjustment policies that direct particular policy and legislative reform (Paarlberg, 2010, Chapter 14;

Phillips & Roberts, this volume). Even though the United States usually dictates such terms rather than being on the receiving end, local schools in the United States are indeed influenced by international and transnational policy. The World Trade Organization, particularly, has tremendous effect on agricultural policy, influencing subsidies and settling disputes over everything from sugar to bananas. Most individuals in the United States, though, never notice these influences.

Also partly outside and partly inside the local system are the environments and structures and the processes that influence the local system. The media is perhaps the most visible and influential of these (to the left of the local box in the figure). One of us (Marcus) has been monitoring Google alerts to media stories on school food (and its various synonyms) from around the world over the past three years, and each day hundreds of stories in English alone appear. There have also been high profile television shows and movies that have brought the politics of food to the fore. Jamie Oliver, the British celebrity chef, has spurred radical changes to school meals in the UK through his reality show, *Jamie's School Dinners* (Gilbert, 2005; see Weaver-Hightower, this volume), and he has garnered tremendous attention and an Emmy award for a similar show in the United States, *Jamie Oliver's Food Revolution* (Seacrest & Smith, 2010). Other films about food, both in school and out, have brought attention to food issues, such as *Food, Inc.* (Kenner, 2008) and *Two Angry Moms* (Kalafa, 2007). High-profile books have also driven concerns, such as those by Eric Schlosser (2001) and Michael Pollan (2006, 2008), and these general concerns about the industrial food system have filtered to and fed from school food reform politics. And of course countless blogs and social media groups have also publicized school food issues, such as Mrs. Q's eco-stunt blog in which she photographed and ate school lunches every day for a year (2009–10; see http://fedupwithschoollunch.blogspot.com). All considered, this media presence for school food politics has provided a fertile ground for reform efforts.

Other dynamics in the U.S. food environment are also acting as mainsprings for school food reform (see the "clouds" in the figure). Concerns over obesity (cf., Obama, 2010; Oliver, 2006; Popkin, 2009), concerns over food safety in schools (covered extensively in the newspaper *U.S.A Today* in 2009 and 2010; e.g., Eisler, Morrison, & DeBarros, 2009), high profile efforts during the Child Nutrition Act reauthorization to raise the reimbursement to schools for lunches served, and concerns over the fate of the U.S. military's preparedness if school lunches don't improve (Mission: Readiness, 2010) have all brought school food politics to the front burner of the U.S. population's attention. Movements for progressive school food reform, in debates over what kids should eat, have struggled against strong currents of personal responsibility ideologies and anti-government sentiment. The culture of industrialized food in the United States—a nationalistic pride in the agricultural and fast food innovations that have transported the "American Way" of life and eating to the rest of the world—combines with the ideologies of personal responsibility and limited government to make criticizing the status quo of food always controversial. Debates over food issues occur in individual schools just as much as in the national forum, so Figure One's single school in North Dakota is densely interconnected with the larger ecology when, for example, teachers and cafeteria workers get together to discuss how to help kids who are overweight by cutting out "junk" food from the cafeteria.

As this analysis illustrates, using an ecology metaphor for school food politics offers a means of moving researchers and practitioners alike toward broader and deeper understandings of how we've come to this point in history and of what the current context looks like. Yet one of the chief benefits of such an approach is that it helps to focus *strategy* (see Robert and Kovalskys, this volume).

> [Ecology analysis] forces one to think about tactics in policy processes in a new, multifocal way. It does so because the metaphor urges expansive thinking, an understanding of interrelationships, and a view of policy as having broad impact.
>
> To strategize in a system that one conceptualizes in this way—to be more successful *because* one has such a perspective—means intervening in the policy process at many points rather than agitating at one particular stage... (Weaver-Hightower, 2008, p. 162)

A policy ecology analysis urges the researcher to identify the influential actors, to understand what relationships exist and must be dealt with or changed, and what challenges reside in the environments and processes of any ecology. These understandings are crucial if changes are to be made in school feeding, as the contributors in Section Two attest.

Why Educators Should Take School Food Seriously

Some readers might still be wondering: Why food? Out of all the issues that consume eductors—including privatization, urban reform, rural desperation, standardized testing, achievement gaps, immigration, girls' participation, second language instruction, and teacher training—it might seem to some a low priority to think about food.

There are clear reasons why food has often been overlooked in education, just as it has been in the other humanities and social sciences. As Belasco (2008, pp. 2–5) notes, academia has often ignored food, partly because it is of the body, an animalistic part of our existence; partly because it is associated with the private sphere of females; and partly because of modern Western societies' beliefs in "technological utopianism," which hides the provenance of our food behind a long chain of reformulation, fortification, and ideological cleansing by advertising and other marketing (on this last point, see also Vileisis, 2008).

All of these reasons are true in educational research, as well, but education also has unique reasons why food has been overlooked. First, learning and the mind are often thought of as separated from the body, so nutrition is thought of as irrelevant to the true mission of schooling. Also, food is often considered utilitarian rather than integral, something that must be done (sometimes grudgingly) when you keep kids for seven or eight hours. Third, school food, if it is seen as vital, is often thought of as purely health oriented, an issue for nutrition rather than academics or socialization. Finally, school food is often overlooked because it is the butt of jokes or the target for vilification (think of popular culture's stereotypic, corpulent "lunch lady" with her hairnet, bad attitude, and exaggeratedly grotesque food); it is not hard to see why educators and researchers would overlook and marginalize a topic of such abjection, leaving it for nutritionists or public health officials to explore.

Despite these reasons for being overlooked, we argue that school food is a crucial topic to consider for educators, policymakers, researchers, and citizens, not one that should be pushed off the radar until all the other problems of education are fixed. Instead, there are specific reasons why educators and researchers should care about food issues in schools, for food issues are interconnected with the other, more obvious issues of social and educational reform.

First among these reasons is that *school food impacts students' health.* The pioneers of school feeding in the late nineteenth and early twentieth centuries advocated for school feeding for health reasons, though they were concerned about malnutrition and a *lack* of calories (Levine, 2008). Now the concern is obesity. U.S. childhood obesity has tripled since 1980, with 9.5% of infants and toddlers and 16.9% of children ages 2 to 19 considered obese (Ogden, Carroll, Curtin, Lamb, & Flegal, 2010). Obesity also has a heavy economic toll, accounting for US$147 billion in health care in 2008, a full 10% of all medical spending (Finkelstein, Trogdon, Cohen, & Dietz, 2009). Yet despite its importance, myopic focus on obesity can erase equally important health considerations, including malnutrition, exposure to pesticides and food additives, allergies, and even the positive contributions of micro- and macronutrients supplied by school meals. For educators, researchers, and policymakers, these issues should be addressed because they can play a big factor in students' abilities to learn and their long-term potential for success.

Similarly, *school food can play an important role in students' academic achievement*. Many studies suggest that *whether*—especially—and *what* students eat influences their attainment of and success in schooling (e.g., Taras, 2005), perhaps even more than it influences health (Hinrichs, 2010). Belot and James (2009), for example, considered the impact of healthier food provision in Greenwich, England, the borough that participated in Jamie Oliver's above mentioned campaign for fresh, whole foods in schools. The authors concluded that Oliver's campaign resulted in substantial improvements in literacy and science tests alongside decreased absenteeism. The School Food Trust (2009), as another example, found that, in English schools that overhauled their food and their dining environments, students were more on-task than control group students. If we are concerned about student learning, food is thus not a distraction but can be an integral factor in improvement.

Third, *school food impacts teaching and administration*. Most obviously, teachers and principals also eat the food, making them subject to health and cognitive impacts, too. Many schools profit from food sales, as well; vending machines and soft drink pouring rights contracts (see Nestle, 2007, Chapter 9) often fund key school programs, and administrators are loathe to give these funds up, whatever the potential health consequences (see also Phillips & Roberts, this volume). Yet school meals provision is also a commitment of time and labor, often adding to educators' existing work inside the classroom, their work outside the classroom advocating for food, or in some cases the work they do serving the food to children (see also Robert and Kovalskys, this volume). One principal that Marcus interviewed, for example, spent her first year as a principal crying nearly every day, not over student learning or teacher development, but over the problems of noise and behavior in the cafeteria. Overlooking food, and thus such pressures, leads us to misunderstand the contexts of schools and those who run them.

The fourth reason to attend to school food is that *students learn about food from schools*. Some of this learning is from the "official" curriculum—family and domestic sciences courses, nutrition lessons, and so on—but much learning about food also comes from the "hidden curriculum" (Jackson, 1968). Using food for manipulatives or art projects (like the iconic macaroni necklace), for example, may teach students to think that food is for play, not valuable, and acceptable to waste. For children with lives marred by hunger, such uses of food are deeply troubling and disrespectful (Hannon, 2006). At the very least, children are taught the value, uses, and customs of food in the cafeteria (see DeLeon, this volume). In the United States, school cafeterias can perhaps best be seen as laboratories for "McDonaldization" (Ritzer, 1993), where efficiency, calculability, predictability, and control of both labor and customers are preeminent concerns. But McDonaldization isn't just a metaphor. About 35.5% of U.S. school districts offer fast food-branded foods (School Nutrition Association, 2009). Domino's pizza is most prevalent, being served in about 26% of all U.S. school districts. In larger districts, urban areas where fast food is already omnipresent outside schools, this increases to 50.5%. And more than the food follows the fast food model. The median time students have for lunch is 25 minutes in elementary and 30 minutes in middle and high schools, though some schools have lunches as short as twenty minutes, barely enough time to *get* food much less savor it. Students, while they rush through their meal, are also subjected daily to a "total advertising environment" in the cafeteria (Molnar, Boninger, Wilkinson, & Fogarty, 2009), with ads pushing energy drinks, candy, the military, milk, and more. While not the same in all cafeterias, massive numbers of American children are being trained to eat fast food or fast-food lookalikes in a fast-food environment. Educators and communities have good reason to pay attention to what children are learning about food (see Lalonde, this volume), for these lessons can follow students throughout their lives.

Fifth, *school food is a window into identity and culture*, important facets of students' lives that are important for educators, researchers, and policymakers to understand. To build on Brillat-Savarin's you-are-what-you-eat quip noted earlier, food establishes who we are in gendered, sexualized, raced, and ethnic senses, and who we are through food has social consequences (Bourdieu, 1984). Ludvigsen and Scott (2009) provide a compelling example. In the English schools they studied, both children and adults saw strict divisions between foods appropriate for children ("junk" food) and those appropriate for adults ("healthy" food); in the children's view, "To eat healthy food was almost viewed as a rejection of the intrinsic meaning of being a child" (pp. 426–427). Perhaps more importantly, the students interviewed made consistent prejudgments about the gender, social class, and identity of people who might consume the kinds of food the interviewers presented; particular foods were for particular *kinds* of people. The children used this acute sense of food and identity, the authors concluded, as a kind of "social camouflage" to fit in or to avoid bullying. Educators would do well to understand such food cultures in schools, for social relations are in part based on these cultures.

School food also impacts the environment and animals. The food distributed in schools often has traveled great distances from farm to plate, using massive amounts of fossil fuels in the process. Combined with the fuel used to refrigerate, freeze, cook, and keep it warm, school food uses vast energy resources, and all has an impact on the environment. Animal production, too, creates enormous environmental impacts, with confined animal feeding operations producing tremendous quantities of methane and manure runoff, not to mention introducing staggering quantities of antibiotics into the food supply and environment (e.g., Canell & Remerowski, 2005; Pollan, 2006). Perhaps more importantly, the mass production of meat can introduce unimaginable cruelties for the animals themselves (e.g., Monson, 2005; Psihoyos, 2009; Simon & Teale, 2009; DeLeon, this volume).

Seventh, school food deserves scrutiny because *it is big business*. According to the SNA (2008), schools in the United States serve nearly seven billion meals each year, including lunch and breakfast. U.S. primary and secondary schools' retail sales equivalent is US$15.9 billion, making it 15% of noncommercial food service or 2.5% of *all* U.S. food service and restaurant sales. Again, though, lunch and breakfast are not the totality of food sold in schools. About 11.5% of the $22.05 billion vending market comes from schools and colleges, as well (Maras, 2009). School stores, canteens, and fundraising–from chocolate bars to bake sales–also generate billions of dollars (Center for Science in the Public Interest, 2007). These giant sums are quite attractive to food manufacturers, of course, particularly since schools are a relatively stable source of income, less subject to the volatility of other food markets; in fact, as one school food broker told Marcus, when the economy goes down *more* kids eat school meals. Schools are also corporations' prime targets for getting children to try new products, view advertising, and develop brand loyalty. Educators, policymakers, and researchers should thus watch carefully what impacts this has on students, schools, and communities.

Finally, *school food impacts social justice*, a crucial reason to move it from the margins to the center of educational thought. Put simply, food practices are a means of social reproduction, oppression, and resistance. As Sandler (this volume) suggests, who is fed, who does the feeding, what is served, where and when, are all questions of importance in looking at school food. The histories of school food in countries around the world, from the United States (Levine, 2008) to England (Berger, 1990) to South Korea (Kang, this volume), are filled with events that pitted social groups against one another and saw struggles for justice and recognition for groups that had been marginalized because of race, gender, social class, and more.

Hunger, of course, remains perhaps the major social justice issue connected to school food globally. Nearly one billion people suffer from food insecurity worldwide (FAO, 2010), and according to the World Food Programme (http://documents.wfp.org/stellent/groups/public/documents/newsroom/wfp199570.pdf), around 66 million children in developing countries attend school hungry daily. Programs to reach these children struggle to provide even rudimentary nutrition, and not because the world produces insufficient calories (e.g., Patel, 2007). One need not travel to Africa or Asia to find hunger, though. The U.S. Department of Agriculture's latest food security report (Nord, Andrews, & Carlson, 2009) estimates that nearly 15% of U.S. households—17 million families—had their access to food limited some time during the year by insufficient financial or other resources. This figure is the highest since statistics were first collected in 1995, and record numbers of children are qualifying for free and reduced-price meals (Eisler & Weise, 2009). This represents a key social justice concern because particular students and communities are bearing the brunt of food insecurity. Clearly, if we are concerned with social justice, food politics must be attended to. For educators, researchers, and policymakers, this requires viewing school food as one of the central facets of school reform alongside welfare policies, transportation policies, health policies, and more (Anyon, 2005). That is precisely how the contributors to this volume view school food.

Overview of the Chapters

Using a broad view of school food as a policy ecology influenced not only the contributors but also our selection of essays for this volume. The contributors speak *from* and *about* various positions within the larger ecology, representing a diverse range of national contexts, professions, reform ideas, and political stances. While some contributors might disagree on fundamental questions—like the ethics of eating meat or the relative roles of government versus individuals in creating reform—all are united in a sense that deep changes are both required and possible. All also speak from a deep knowledge of the complexities of the ecologies they work within.

The authors in Section One look qualitatively at various international contexts, showing the intricacies of the locations and their struggles with school food reforms, often through cases that extrapolate to a national scale. In Chapter One, Jen Sandler lays out the core social justice question at the heart of critical school food reform. This is, in her words, "*Who feeds whom what, how, and for what purpose?*" She then breaks this question down, piece by piece, as a means of examining the role of after-school feeding programs in the urban United States. Examining specifically the work of an organization she calls the Coalition for Local Initiatives, Sandler is able to unsettle dominant discourses of the deficient, poor urban parent, taking us instead inside the struggles communities wage to provide healthy and just food. She does so to ask us to "pay attention to the details of reform, and to be aware of how reform agendas to respond to conditions of urban poverty shape our ideas about the urban poor."

In Chapter Two, Marcus Weaver-Hightower examines cases of school food reform organizations in three countries: Australia, England, and the United States. He first gives the history of each country's school meals service and then turns to explore the cases of, respectively, the Healthy Kids Association, the School Food Trust, and the Physicians Committee for Responsible Medicine. These cases, when synthesized, show the similarly tenuous and highly political nature of trying to reform school food across these Anglophone countries. Common themes include the need to think ecologically, the centrality of money, the importance of marketing and research, and the need to work cooperatively with industry.

In Chapter Three, Kristin Phillips and Daniel Roberts present their historical and ethnographic work on school cultivation in the East African nation of Tanzania. While in current discourses the school garden is often presented as a positive force to increase food knowledge and healthy eating (e.g., Waters, 2008), Phillips and Roberts remind us that in developing countries gardening and farming can sometimes be exploitative, too, with the labor of students used for the gain of school personnel. They urge readers and international aid organizations to reexamine notions of food sovereignty and food security, suggesting that these discourses overlook many of the dynamics that create hunger in developing countries.

In Chapter Four, Sarah Robert and Irina Kovalskys focus on a vexing problem in Argentina (one that many societies face): the simultaneous occurrence of hunger and obesity within groups facing poverty and food insecurity. The authors create dialogue between the qualitative understandings of educators in schools (Robert) and the quantitative understandings of public health professionals working to establish the epidemiology around food and nutrition (Kovalskys). Often, they say, these two groups work at cross purposes, as if these problems were unconnected. They call for urgent cooperation between groups who are ultimately working to the same end. Cooperation of actors is all the more urgent, they argue, in the wake of economic crisis.

In Chapter Five, Mi Ok Kang provides a detailed history of the politics of providing universally free school meals to students in South Korea. She traces the movements and discourses that have worked through electoral politics in the 1990s and 2000s. It is a story that incorporates complexities of finance, competing ideologies of government's roles, food safety scares, struggles by citizens and politicians alike, and the tense interconnections between levels of administration. In the end, Kang shows a hopeful and progressive result in recent elections that have put universally free meals within reach.

Section Two of the collection moves away from national contexts and organizations, focusing instead on local contexts or on major ideas that have defined recent school food reform. Chapters in this section provide theoretical explorations of how things have come to be, rationales for fundamental changes, and, often, firsthand accounts from those who have been on the front lines of making reform happen.

In Chapter Six, Marion Nestle makes a persuasive case for why school food should be a "hot button issue." Nestle's focus is on the precarious position of the United States Department of Agriculture, caught between the students they feed, the farmers they are tasked to serve, and the industries that revolt at any suggestion that students "eat less" of anything. She argues that the public's and the government's focus needs to shift, assuring children a space in schools where "their needs come first—not their future as consumers."

In Chapter Seven, Catherine Lalonde describes her challenges in creating a cooking class for African American and Latino teenagers at an urban community's youth center, which she calls Prospect Clubhouse. Lalonde outlines her program, based on critical pedagogy tenets, which sought to teach the youth at the club both cooking skills and new ways of thinking about the messages they encounter about food. Hers is a venture that sought to interrupt the increasing reliance on processed, grab-and-go food, a reliance that deskills communities to the detriment of their health.

In Chapter Eight, Doug Davis, Dana Hudson, and members of the Burlington School Food Project outline the multifaceted program that Burlington, Vermont, public schools have implemented to incorporate more fresh, locally produced foods. They describe the many ways they have sought to accomplish this, including farm-to-school initiatives, a community supported bakery program, and involving students and community members in competitions, note card sales, and volunteering opportunities. Davis and coauthors lay out the ups and downs of all this work, showing concretely how success is achievable given effort, cooperation, and the proper supports.

In Chapter Nine, Abraham DeLeon gives readers a postmodern examination of the school cafeteria, particularly its role in encouraging the mass consumption of meat. Taking an animal liberationist perspective, DeLeon urges a rethinking of the spatial practices that reproduce the oppression of nonhuman animals, a dynamic—which he likens to "fascism"—that ultimately supports other forms of oppression among humans (colonialism, racism, sexism, and more). He poses a response to this oppression based in anarchism, "relying on communities that subscribe to some sort of communal or social justice paradigm to solve social problems for themselves" rather than relying on the state to act to end the oppression.

In Chapter Ten, our Coda, we editors revisit the aims of the compilation and synthesize the major themes that we have come to understand from the cases the other chapters present. These understandings urge us all, we argue, toward healthier horizons in school food.

We also include an appendix of suggested websites providing research, resources, recipes, and curricula on school food. These are intended as supplemental information for policymakers, educators, child nutrition professionals, and citizens who want to make reforms toward healthier, more just food provision in schools.

It is our fervent hope that, taken together, the chapters in this volume illuminate for readers the complexities of school food and the politics that seem to perennially swirl around it. In putting this collection together, we have tried to show the diversities of feeding schemes and their challenges. It is easy sometimes, in the face of complexity and interdependency, to feel overwhelmed, as if there is too much to do and too many layers to fight against. Rather than the urge toward paralysis, though, we hope that readers will find some inspiration and guidance from the many cases presented in the chapters here. The contributors show just a few of the many ways that committed actors across these varied ecologies have bravely and creatively tackled needed reforms. We hope that their work energizes and informs readers because, in the end, it is social justice and human rights work to provide nourishment and education to the children of the world.

References

Anyon, J. (2005). What counts as educational policy? Notes toward a new paradigm. *Harvard Educational Review*, 75(1), 65–88.

Belasco, W. (2008). *Food: The key concepts*. Oxford, England: Berg.

Belot, M., & James, J. (2009). *Healthy school meals and educational outcomes* (Report No. 2009-01). Essex, England.

Berger, N. (1990). *The school meals service: From its beginnings to the present day*. Plymouth, England: Northcotte House.

Bourdieu, P. (1977). *Outline of a theory of practice* (R. Nice, Trans.). Cambridge, England: Cambridge University Press.

Bourdieu, P. (1984). *Distinction: A social critique of the judgment of taste* (R. Nice, Trans.). Cambridge, MA.: Harvard University Press.

Bourdieu, P., & Passeron, J. C. (1990). *Reproduction in education, society, and culture* (R. Nice, Trans. 1990 ed.). London: Sage.

Canell, M., & Remerowski, T. (Directors). (2005). *Frankensteer* [Motion picture]. Canada: Bullfrog Films.

Center for Science in the Public Interest. (2007). *Sweet deals: School fundraising can be healthy and profitable*. Washington, D. C. Retrieved from http://www.csp net.org/schoolfundras ng.pdf

Eisler, P., Morrison, B., & DeBarros, A. (2009, December 9). Fast-food standards for meat top those for school lunches, *U.S.A Today*. Retrieved from http://www.usatoday.com/news/education/2009-12-08-school-lunch-standards_N.htm

Eisler, P., & Weise, E. (2009, June 11). More students on free lunch programs, *U.S.A Today*. Retrieved from http://www.usatoday.com/news/education/2009-06-10-student-lunches_N.htm?csp=34

Finkelstein, E. A., Trogdon, J. G., Cohen, J. W., & Dietz, W. (2009). Annual medical spending attributable to obesity: Payer- and service-specific estimates. *Health Affairs, 28*(5), w822–831. doi: 10.1377/hlthaff.28.5.w822

Food and Agriculture Organization. (2010). *The state of food insecurity in the world: Addressing food insecurity in protracted crises*. Rome, Italy. Retrieved from http://www.fao.org/docrep/013/i1683e/i1683e.pdf

Gee, J. P. (2005). *An introduction to discourse analysis: Theory and method* (2nd ed.). New York: Routledge.

Gilbert, G. (Director). (2005). *Jamie's school dinners* [Television series]. England: Freemantle Media.

Hannon, J. (2006). Lessons from Ana. *Rethinking Schools, 20*, 47.

Hinrichs, P. (2010). The effects of the National School Lunch Program on education and health. *Journal of Policy Analysis and Management, 29*(3), 479–505. doi: 10.1002/pam.20506

Imhoff, D. (2007). *Food fight: The citizen's guide to a food and farm bill*. Healdsburg, CA: Watershed Media.

Jackson, P. W. (1968). *Life in classrooms*. New York: Holt.

Kalafa, A. (Producer & Director). (2007). *Two angry moms* [Motion picture]. United States: A-RAY Productions.

Kenner, R. (Director). (2008). *Food, Inc.* [Motion picture]. United States: Participant Media.

Levine, S. (2008). *School lunch politics: The surprising history of America's favorite welfare program*. Princeton, NJ: Princeton University Press.

Lien, M. E., & Nerlich, B. (Eds.). (2004). *The politics of food*. Oxford, England: Berg.

Ludvigsen, A., & Scott, S. (2009). Real kids don't eat quiche: What food means to children. *Food, Culture & Society, 12*(4), 417–436. doi: 10.2752/175174409X456728

Maras, E. (2009, August). Recession drives profit protection initiatives. *Automatic Merchandiser, 51*, 28–42.

Mission: Readiness. (2010). *Too fat to fight: Retired military leaders want junk food out of America's schools*. Washington, D. C. Retrieved from http://cdn.missionreadiness.org/MR_Too_Fat_to_Fight-1.pdf

Molnar, A., Boninger, F., Wilkinson, G., & Fogarty, J. (2009). *Click: The twelfth annual report on schoolhouse commercialism trends: 2008–2009*. Boulder, CO. Retrieved from http://epicpolicy.org/publication/Schoolhouse-commercialism-2009

Monson, S. (Director). (2005). *Earthlings* [Motion picture]. United States: Nation Earth.

Nestle, M. (2007). *Food politics: How the food industry influences nutrition and health* (Revised and expanded ed.). Berkeley: University of California Press.

Nord, M., Andrews, M., & Carlson, S. (2009). *Household food security in the United States, 2008* (Report No. 66). Washington, DC

Obama, M. (2010, March 14). Michelle on a mission: How we can empower parents, schools, and the community to battle childhood obesity. *Newsweek, 155*, 40–41.

Ogden, C. L., Carroll, M. D., Curtin, L. R., Lamb, M. M., & Flegal, K. M. (2010). Prevalence of High Body Mass Index in U.S. Children and Adolescents, 2007–2008. *Journal of the American Medical Association, 303*(3), 242–249. doi: 10.1001/jama.2009.2012

Oliver, J. E. (2006). *Fat politics: The real story behind America's obesity epidemic*. New York, NY: Oxford University Press.

Paarlberg, R. (2010). *Food politics: What everyone needs to know*. Oxford, England: Oxford University Press.

Patel, R. (2007). *Stuffed and starved: Markets, power, and the hidden battle for the world food system*. London, England: Portobello Books.

Pollan, M. (2006). *The omnivore's dilemma: A natural history of four meals*. New York: Penguin Books.

Pollan, M. (2008). *In defense of food: An eater's manifesto*. New York: Penguin Press.

Popkin, B. (2009). *The world is fat: The fads, trends, policies, and products that are fattening the human race*. New York, NY: Avery.

Poppendieck, J. (2010). *Free for all: Fixing school food in America*. Berkeley: University of California Press.

Psihoyos, L. (Director). (2009). *The cove* [Motion picture]. United States: Lionsgate.

Ritzer, G. (1993). *The McDonaldization of society: An investigation into the changing character of contemporary social life*. Thousand Oaks, CA: Pine Forge Press.

Schlosser, E. (2001). *Fast food nation: The dark side of the all-American meal*. Boston: Houghton Mifflin Company.

School Food Trust. (2009). *School lunch and learning behaviour in primary schools: An intervention study*. Retrieved from http://www.schoolfoodtrust.org.uk/partners/reports/school-lunch-and-learning-behaviour-in-primary-schools-an-intervention-study

School Nutrition Association. (2008). *Little big fact book: The essential guide to school nutrition*. Alexandria, VA: School Nutrition Association.

School Nutrition Association. (2009). *School nutrition operations report: The state of school nutrition 2009*. National Harbor, MD.

Seacrest, R. (Producer) & Smith, B. (Director). (2010). *Jamie Oliver's food revolution* [Television series]. United States: American Broadcasting Corporation (ABC).

Shear, L., Means, B., Mitchell, K., House, A., Gorges, T., Joshi, A. et al. (2008). Contrasting paths to small-school reform: Results of a 5-year evaluation of the Bill & Melinda Gates Foundation's National High Schools Initiative. *Teachers College Record, 110*(9), 1986–2039.

Simon, T., & Teale, S. (Directors). (2009). *Death on a factory farm* [Motion picture]. United States: HBO Documentaries.

Taras, H. (2005). Nutrition and student performance at school. *Journal of School Health, 75*(6), 199–213. doi: 10.1111/j.1746-1561.2005.tb06674.x

Vileisis, A. (2008). *Kitchen literacy: How we lost knowledge of where food comes from and why we need to get it back*. Washington, DC: Island Press.

Waters, A. (2008). *Edible Schoolyard: A universal idea*. San Francisco: Chronicle Books.

Weaver-Hightower, M. B. (2008). An ecology metaphor for educational policy analysis: A call to complexity. *Educational Researcher, 37*(3), 153–167. doi: 10.3102/0013189X08318050

• SECTION ONE •

From Pap to Sloppy Joes to Nada

Inside International School Food Politics

• CHAPTER ONE •

Reframing the Politics of Urban Feeding in U.S. Public Schools

Parents, Programs, Activists, and the State

Jen Sandler

School Feeding in Conditions of Urban Poverty: The Landscape

My contention is that an understanding of school feeding in contemporary conditions of urban poverty in the United States requires a view of public policy through the lens of profoundly local negotiations, contextualized within a broad account of historical struggles. When a public school feeds, when a business feeds, or when a charity or nonprofit organization feeds, the dynamics of this feeding—who feeds whom, what, how, and for what purpose—say a great deal about how children are cared for by society. Debates over inner-city school feeding in the United States have historically focused on *whether* to feed hungry children living in poverty—on whether, ultimately, as a society "we" care about "them" at all. Current school food debates focus almost exclusively on *what* to feed low-income children. The subtext of these contemporary feeding-curriculum debates, which presume that "we" professionals and policymakers care and that, in some way, their parents do not or cannot care for them properly, fail to take into account the structural and political shifts in feeding, as well as the contemporary dynamics of a great deal of school feeding in conditions of urban poverty. The discourses of uncaring parents and debates over how and whether to regulate what is fed distort and ignore the complex historical and contemporary dynamics between families, schools, private organizations, and food that take place across urban areas in the United States.

This chapter calls for a reframing of the debate about school feeding in urban centers of the United States, focusing on both the local negotiations of school feeding and the critical-structural context for these negotiations. Before embarking on this reframing, it is necessary to say something about the current frameworks for thinking about the politics of school food. These frameworks are characterized by (1) dominant popular discourses, (2) polarized policy discourses, and (3) a lack of attention to the actual structure of school feeding in U.S. inner cities, which, put simply, extends far beyond lunch. I will briefly describe these barriers to both deeper and broader insight into contemporary urban school feeding.

First, there are the dominant popular discourses. From the left as well as the right, and from Jamie Oliver to *The Biggest Loser*, dominant discourses individualize responsibility for poor health. These discourses maintain that the poor, inner-city parent—usually single, usually of color—is a deficient parent: they do not know—or, in the strongest form of this discourse, they do not care—about their children's health and well-being. They cannot be trusted to make good decisions for their children, so these parents must be either taught or compensated for by state actors. Urban schools must step up and compensate for what children are not getting at home by providing nutritious meals and teaching children decision-making skills. That, in fact, is the job of the inner-city school in this model—to compensate for poor homes, poor eating, and poor parenting. The way to do this, according to dominant discourses of poverty, is not to combat poverty itself, but to either teach parents to be good parents or to institute policies that make up for the deficits in a child's home.

Second, there is debate about what policies should address these problems. In school feeding, the main publicized form of this debate has taken place around the (substantial) component of school feeding that is not federally regulated. I refer here to the sale of food to children through vending machines, school stores, a la carte lines, and other private sources of "subsidizing" federally subsidized school food programs (as well as the coffers of school districts). Debate about supplementary school food policies takes place at every level, from school district to state and federal legislative bodies. On one side, such debates generally invoke the paternalistic role of schools in protecting children from ignorant or indifferent poor parents. On the other side of the political spectrum, these debates invoke the importance of choice and personal responsibility. These two sides are visible in myriad recent public policy debates and congressional hearings on school food policy (e.g., United States Congressional Committee on Agriculture, Nutrition, and Forestry, 2010). These are the two poles of the public policy debate: do we make poor parents responsible for changing their children's behavior and eating the healthy school lunch rather than the packaged frosted cupcakes, or do we declare them hopelessly incompetent and use policy to instruct children—and by extension parents—in proper eating?

Finally, there is the structure of urban school services, which in some ways lessens the relevance of many of the above public policy debates. The explosion over the past twenty years of contracting out food services to private providers, specifically, is part of a broader move toward privatization and contracting of *all* school services, including the provision of comprehensive tutoring, recreation, health, child care, and more in urban schools. Before- and after-school programs, in particular, have become a staple component of urban districts. Funded by an irregular mix of federal, state, district, and philanthropic dollars, along with some contribution from the families that participate, before- and after-school programs often take responsibility for the care and feeding of children for two hours in the morning and for another two to three hours in the afternoon. This feeding by contracted private service providers, much of which takes place outside the National School Lunch Program and the School Breakfast Program, occurs with little or no oversight or regulation. This has resulted in an unregulated and unsystematic system in which much inner-city children's learning, eating, and recreating at school is not controlled by or accountable to the school itself (or, by extension, to the broader system of public governance). Surreptitious privatization of urban public schooling has changed the circumstances of urban school feeding just as it has changed the circumstances of urban school learning. These changes are no less real for having gone largely unacknowledged in the public sphere.

These three layers of school food politics—discourses, policies, and the structure of services—do not cohere. We have discourses of ignorant poor parents (usually single women of color) within which policy advocates and program professionals argue over whether to compensate for poor parents' failures through stricter policies, on the one hand, or whether to, on the other, endow these parents with more choices and hold them responsible for the consequences. Add to this the actual landscape of urban schools in which much of low-income children's hunger is being addressed by hundreds of partially publicly funded private organizations that engage with the dynamics of feeding as they see fit.

In such complex circumstances, ethnographers are wont to move both closer in and further out, to gain a thick understanding of day-to-day life in urban schools from the perspective of those targeted by policies and reforms, and to also place this day-to-day life in its broader political context and historical perspective. This chapter thus centers on quite different key figures than those emphasized in the discourses above. My key figures are (1) parents who attempt actively to negotiate their children's school and after-school eating; (2)

private (non-school-run) urban after-school programs through which much feeding and negotiating takes place; (3) the state, in the form of both local professionals and federal policymakers, as seen over time; and (4) independent activist groups of the past, like the Black Panther Party, that have interfaced with the state over the politics of school feeding. Focusing on these figures suggests a different story of the politics of school feeding than that of the past, a story in which the question is not "federal versus local" or "choice versus health," but instead the critical question: *who feeds whom what, how, and for what purpose?*

Localizing Urban School Feeding: A Single Event in Context

The event that grounds this conceptual reorientation could not possibly be more local. I examine a typical early evening meeting of a school's site council, a group of parents and community members who governed the community services provided at a public school in a low-income urban neighborhood of Crossroads (this and all other names are pseudonyms). Crossroads is a large and diverse metropolitan area in the midwestern United States that served as the location of my ethnographic fieldwork in 2006. I chose Crossroads in order to study its Coalition for Local Initiatives, a large non-profit organization that uses community organizing and local governance mechanisms to guide social services and policy that affect low-income families. The site council is the Coalition's main instrument of neighborhood-school governance, and over the course of eight months I attended more than 100 Coalition meetings involving site council members. The main criterion for site councils is that they cannot be led by or controlled by professionals—people who are paid to provide services, such as teachers, counselors, social workers, and administrators. Rather, site councils are supposed to be volunteer-led governing bodies, comprised of parents and neighbors who have a stake in what goes on at the school.

The activities of the sites include a wide variety of neighborhood institutions, including health and dental clinics, job fairs and adult education, social services, prevention programs, and cultural and civic activities. By far the most ubiquitous and prominent activities governed by site councils—and provided by professionals hired by the Coalition for Local Initiatives—are before- and after-school programs. Site activities, including before- and after-school programs, are funded through a variety of district, state, and federal funding streams. For example, in 2006 some sites were financed through 21st Century Community Learning Centers, a $1.5 billion federal funding stream. Others were funded through the state's Department of Social Services. Government money, in some form or another, supports more than 90% of before- and after-school program activities. I was working with the Coalition to study its local governance processes, so I attended many meetings where the agenda focused on leadership, policy, and services. Food served as a backdrop to virtually all of these meetings, with dinner or snacks included for parents and their children, and food often became a subject of discussion.

Most importantly, food was a significant component of all of the Coalition's before- and after-school programs. These programs allowed children to be dropped off at school as early as 6:30 A.M., and to be picked up at 6:00 P.M. (or even later on days when there was a site council meeting). The Coalition for Local Initiatives was thus in charge of thousands of Crossroads' low-income children's care—and often feeding—for between one-third and one-half of the nearly twelve hours that many children spent at school. This is not unique to Crossroads. Throughout the inner cities of the United States, many thousands of poor children are fed, educated, exercised, and entertained before and after the official school day, often at their schools, by private organizations using mostly public (although increasingly also philanthropic) dollars. The school cafeteria, in short, is not the only context in which public dollars feed children at school. For many inner-city children, contracted independent programs provide not only tutoring, child care, and recreational activities; they also provide nutrition in the forms of snacks, some breakfasts, and, increasingly, some form of an evening meal.

What systematic knowledge do we have about publicly funded feeding of children by non-public organizations, like the Coalition for Local Initiatives, in U.S. inner cities? Very little, it turns out. Although there is official regulation of food paid for by the United States Department of Agriculture (USDA), there is no systematic regulation or oversight of after-school feeding specifically. Oddly, in light of recent interest in school nutrition, feeding through contracted after-school programs has barely registered on the public radar screen. For example, in the $1.5 billion federal funding stream for 21st Century Community Learning Centers, "nutritious snacks" are listed as eligible for contract dollar expenditures, but there is no reliable federal mechanism for or interest in interrogating food receipts, and no requirement that they use food vetted for compliance with USDA school meal regulations. When I spoke with several administrators of federal after-school program funding streams about how they make decisions regarding which local programs to fund, not surprisingly, none of them mentioned food. Feeding is subtext rather than text, presumed to be important enough to include in a contract, but its details—unlike school lunches—not considered a matter of government concern. Urban feeding of poor children through contracted services is thus decentralized and, based on its absence as a topic of discourse in public policy and even by school-food activists, off the national radar screen. Food is simply a minor category of allowable expenses under large government funding streams. But feeding is certainly a hot topic for the many hundreds of thousands—if not millions—of children, parents, and professionals involved in urban after-school programs in the United States.

In the meeting I am focusing on from one of the Coalition for Local Initiatives' site councils in 2006, food dominated the conversation. The parents were planning a field day for the school, as the school's principal had deemed the once-annual Field Day beyond the reach of the school's budget. The seven parents and their thirteen children, the latter having been in and around the spotlessly clean but irredeemably shabby elementary school building for twelve hours by this time, were eating sandwiches and juice in the trailer annex that houses the before- and after-school program (as well as child care on meeting nights like this one). The six mothers and one father—five African Americans and two Latinas—had come together on this night under a great deal of stress. Their particular site council imploded the year before due to a combination of toxic leadership and culturally insensitive Coalition staff. Still, the Coalition then hired Rachel, a young and naïve white woman, to take over as the new professional site coordinator. James, an African American Coalition community organizer, was coaching Rachel and overseeing the rebuilding of this site council. The parents on the site council began rebuilding, starting with small tasks, working collectively to make something happen for their kids. They were starting small—seven parents, Rachel, and James—by planning Field Day.

The Field Day discussion was all about food, so much so that it took me some time to figure out what the activities were. Mona, who had clashed with the last site coordinator, declared early in the meeting that her little boy was sick of the food they served him in the program, and she wanted the kids to have something good to eat for once. Rachel, the young, white site coordinator, blushed and stammered that they try to make the food healthy but include some sweet things for snacks, too, and that Mona's son had not complained to her. The room felt tense. James, the Coalition community organizer, redirected the conversation, quickly noting that the parents could have a say in the food and that Field Day was a good place to start. He asked Mona what her son likes to eat. She said hot dogs, but not the overcooked, rubbery kind the school cafeteria serves. She seemed to be redirecting her earlier criticism of Rachel's program to a criticism of the school.

James then asked Rachel if it isn't true that the parents can buy whatever food they want for Field Day and cook it however they want. Rachel, seemingly relieved to have something positive to say, said that yes, they can get whatever they want. She explained the reimbursement process, how they have contracts with some places so parents would not even need to lay out their own money. Another parent, a friend of Mona's, interjected.

> Well, anything? Y'all know we had a problem last year because [the Coalition] said they would reimburse and then they said there were regulations and went back on their word that we could do whatever we want, so you know I just want to know what the rules are to be sure. I mean who gets to decide if something is a special treat or if it's too sweet or what? Can we buy candy? That's not all I'm sayin' because I don't even want to, necessarily, but you know that's maybe where we should start.

She said this much more like a declaration—even an accusation—than a question. I later found out the background to this interjection was a protracted, months-long argument a year prior between parents and Coalition senior staff who had created a "no candy" policy and then came under attack when a parent bought $75 worth of candy for an event and the Coalition refused to reimburse her. This argument became a proxy for cultural disputes between black site council parents and white Coalition site staff, and was addressed—but obviously not truly resolved—by a black Coalition executive.

The tension was rising in the room at this point. But then a parent who had been quiet—and who I later learned was new to the site council—piped up and said,

> I was wondering. Maybe I could buy some ingredients? Maybe make something like with beans and *arroz* [rice], some cheese, maybe even some special meal I know my daughter likes, and her friends? I was thinking to cook would be best if there is help with the money. ... Can we just use ingredients? Is it allowed?

James jumped in and said yes, of course, this was more than just allowed—that would be great! "Anything, anything like that! That sounds delicious," he said. The tension seemed to have been broken, and the parents suddenly talked about what they would make, who would do the shopping, and where they would go. There was some concern to communicate to the children that they shouldn't eat the school lunch because Field Day would start right after lunch, and some parents wanted to make sure the kids would be hungry for the good food. Another parent said, "That's silly! My kid can eat ten hot dogs no matter what he's already eaten, and by the way," she said, "I know a local storeowner who would donate hot dogs, and I'll be at Field Day to remind my son he can't eat all of them himself."

By now everyone was laughing and talking about the food, whether to request time off work or to just cook the food and drop it off in the morning, who would be able to be there and how those who need to work would get the food to them, and other such logistics. Soon a staff person stuck her head in the door to say that the kids were getting really restless. The mood was positively jovial by then, and the parent who was so critical earlier said lightly, "Oh, all right honey, we were just closing up here!" Rachel said everyone should take food home; the plate of cookies, cheese, crackers, and other snack foods she had meticulously set up by the door remained mostly untouched. One of the women said no, that she was watching her calories and cheese was going to be her downfall, and another said she would take some home for her boys, that they would eat anything even though they don't much like carrots. Rachel looked a little insulted, or perhaps just disappointed, but James quickly pulled her into a conversation with one of the parents. Everyone mingled and, within a few minutes, jostled out the door to pick up their children and head home.

This event was not unusual. It was particular, to be sure. But events like this one, in which parents' varied attempts to contribute to their children's and their community's well-being rub up against private institutional structures, schools' resource constraints, intensely local conflicts, and the daily struggles of urban poverty, occur in many different settings on any given night in low-income urban U.S. neighborhoods. Of course, such meetings do not always focus on food. But food—just as hunger and nutrition-related disease—is virtually always in the background, woven throughout the urban landscape of family-school interactions that are mediated by contracted, school-based urban after-school and family support programs.

Toward Analysis

How do we understand the politics of urban school-based feeding in such irregular circumstances? How do we get beyond the discourses about ignorant and indifferent poor parents—caricatures that do not accurately represent the realities of urban families—to look at how parents and children negotiate issues of food in contemporary urban school contexts? How do we get beyond the policy debates that revolve around questions of paternalism versus choice, and which do not generally address or even acknowledge the many—and expanding—spaces of urban school feeding that are structurally untouched by school policy? How do we begin to think seriously about the varied, underchronicled efforts of families living in urban poverty to negotiate diverse local circumstances of school feeding?

In order to think through these questions, I felt a need for guiding questions that cut through the ways school food politics are commonly framed, guiding questions that recognize a broader context than current tightly circumscribed debates. I find these resources in the questions that have guided the development of critical studies of education generally. Indeed, critical scholars of schooling—of teaching and learning—have learned to re-direct us toward broader questions, questions that force us to place current educational reform debates into broad historical, political, and socio-cultural context.

Just as school teaching has never been an apolitical act of educators providing knowledge to ignorant children, school feeding has never simply been an apolitical act of educators providing food for hungry children. Schools are institutions that have always had political, economic, social, and cultural implications. School teaching and learning is and has always been contentious, and any critical inquiry of schooling entails highly political questions: Who should teach? How should children be taught? What should be taught? And what is the function or purpose of schooling for the current society? The critical question for schooling in any given circumstance can be summarized as: Who teaches what to whom, how, and for what purpose? Likewise, a critical inquiry of school *feeding* entails the same type of question:

"Who feeds whom what, how, and for what purpose?"

This question constitutes a grammar of critical inquiry that makes it possible to analyze the complexities of contemporary school feeding, to compare school food politics across contexts, and to investigate the history of these politics. In the following sections, I break down this critical question to trace each of these shifts, building a structural narrative of both the expansion and the politics of urban school-based feeding in the past century that suggests how we might better contextualize, analyze, and understand the significance of the dynamics of feeding in urban schools—during lunch and beyond—such as those in the school site council meeting above.

Who Feeds? Who Is Fed?

Before the National School Lunch Program, school meals were a purely local affair. In the early Twentieth century, poor children were fed by charitable efforts organized locally by religious or civic initiatives (Levine, 2008). Local government-funded school feeding began with the expansion of dire poverty during the Great Depression, then expanded to a federal program in 1935 as a means to address agricultural surpluses. It was contracted in World War II and finally became codified in the National School Lunch Act of 1946 (Levine, 2008). The Act stipulated that states receiving aid would feed free and reduced-price lunches to children who could not afford to pay. At the beginning, the program's allocations more than doubled participation in school lunch programs (Demas, 2000).

In the post-war decades the suburbs expanded, civil rights struggles began, and white flight reconfigured urban landscapes in cities from Chicago and Detroit to Hartford, Atlanta, and Kansas City. School resources followed property values—which, in turn, followed whites—out of the inner cities. Entitlement programs like the National School Lunch Program failed to meet the needs of both the persistently segregated urban and rural South as well as the newly re-segregated urban North. The National School Lunch Program, a federal program designed to address the dual needs of the hungry individual and big-agricultural "victims" of economic collapse in the wake of the Great Depression, simply did not adjust to the new urban landscape. Both the white suburban middle class and the black and Latino inner-city poor populations swelled, and the latter went largely unnoticed in popular culture and federal political consciousness. The new face of urban poverty was pushed aside, ignored, and denied until the mid-1960s.

The struggle that best captures the politics of school feeding in the late 1960s is the fight over breakfast. In 1966, at the same time the Child Nutrition Act made free and reduced-price lunch for all poor children a funded federal entitlement program, the federal government initiated a two-year pilot school breakfast program. The pilot program served over 500,000 children nationally and experienced high demand for expansion. It was unfocused, however, not targeted to particular districts or populations. Due to the vicissitudes of the late 1960s political climate, the federal school breakfast program waxed and waned dramatically after its pilot years, but did not expand from its initial size.

What happened between 1969 and 1971—during a period when the federal breakfast program numbers waned—is an important part of the story behind why urban school feeding expanded. This was the height of the Civil Rights Movement, and militant activist groups were flexing their muscles and demanding change in cities across the country. The Black Panther Party is one of the most famous militant organizations of this time period. Its original agenda consisted of two main urban community activities: survival programs and protection programs. As Marine (1969) explained, the "Party was a revolutionary party and...its immediate purposes are not only to *protect* the ghetto but to *serve* it, and to demonstrate to other black Americans that they mean what they say" (p. 74). The bedrock "survival" program of the Party was called Free Breakfast for School Children. Through this program, the Black Panthers—usually in cooperation with urban black churches—fed an estimated 45,000 children daily in 45 cities nationwide beginning in 1969 (Churchill, 2001, p. 87).

The U.S. Federal Bureau of Investigation (FBI)—under J. Edgar Hoover's orders—worked hard in 1970 and 1971 to undermine the Black Panthers' breakfast program because they saw it as the most subversive of the Panther party's "survival" programs. In essence, the FBI was concerned about the Party's survival programs for precisely the reason the Party initiated them; these programs illustrated the government's weakness and gained sympathizers to the militants' cause. The covert (at the time) agenda to undermine the survival programs was explicit, as documented through a series of FBI memos (cited in Churchill, 2001, pp. 87–89, 259). FBI agents spread the news in some cities that the breakfast was poisoned. In other cities they sent letters to the churches from fabricated "concerned citizens" questioning the political intentions of the breakfast programs and their appropriateness in the neighborhoods. Through violence and propaganda, FBI efforts eventually destabilized the Black Panther Party leadership and organization, reducing and ultimately shutting down the survival programs. Once the Party's survival programs were undermined, Congress began to move school breakfast toward an expanded entitlement program targeted precisely at those populations that the militant civil rights activists had engaged and involved. In 1971 the school breakfast program was funded more heavily again and with a new focus—in policy, at least—on "the neediest of schools." The program became permanent in 1975 and has grown steadily since, but always modestly in relation to the National School Lunch Program. The breakfast program served 9 million children per year in 2008 (more than 30 million were served lunch), at a federal cost of approximately $2 billion, about a quarter of what is spent for school lunches (School Nutrition Association, 2009).

The struggle over breakfast in the late 1960s and early 1970s reveals a new type of struggle over who will feed whom. The larger story was played out, of course, over lunch. In response to urban political activism and social unrest as well as to a broader anti-hunger movement in the 1960s (see Poppendieck, 2010, pp. 58–64), the 1970s was a decade of vast expansion in public programs addressing the most immediate basic needs of the urban poor in the United States. When the National School Lunch Program was made more robust following congressional inquiries and hearings in 1964–65, urban schools (many of them by this time in the midst of massive desegregation efforts) began to claim an earnest intent to feed all hungry children. The expansion of federal Child Nutrition programs lasted through the Johnson, Nixon, Ford, and Carter administrations. During the 1980s, however, the Reagan administration attacked "overcertification" and alleged fraud, cutting subsidies to "non-needy" children. These cuts increased the cost of food for many once-eligible children, reducing participation in the program by 3 million (Levine,

2008, pp. 174–175; for a broader discussion of this era, see also Katz, 1989). Since then, despite entitlement programs that have been cut deeper in each administration, school feeding programs have never earnestly come up on the congressional chopping block (Poppendieck, 2010, pp. 204–205). Today, when children do not eat a school-staff provided breakfast and lunch, it is usually not because it is unavailable to them (though not all schools participate in the NSLP), but because they are eating food from school vending machines, snack bars, contracted private food providers, or supplementary programs (Levine, 2008, Epilogue). Who is feeding has shifted, in other words, from public schools themselves toward private, for-profit contracted food providers along with non-profit community programs.

Today, then, who is fed? The trajectory of more feeding for more children has not substantively eased after the expansions of the late 1960s and 1970s, despite a sizeable dip resulting from the Reagan administration's reforms. The expansion of school feeding programs beyond lunch has in fact continued, with before- and after-school programs in over half of urban districts, as well as summer programs in more than 80% of large urban district schools (School Nutrition Association, 2009). This expansion has come in the form of competitive contracts rather than universal, standardized entitlement. Nevertheless, more inner city children are receiving a higher percentage of their daily calories from services provided at—though not necessarily *by*—the school than ever before.

What Is Fed?

Whereas in the 1960s who was fed was the primary concern of activists, today what is fed in schools is the main topic of U.S. public and political debate over school feeding. Indeed, the nutritional value, provenance, and quality of what children are fed in schools have become so central to contemporary debate that I will limit my discussion of it here.

What historians of the National School Lunch Program have documented thoroughly is that the content of school-provided food is forged from the intertwined politics of agricultural surplus, USDA regulations, cooking facilities limitations, and (more recently) packaged and fast-food corporate lobbying on various fronts for the huge, captive market of public school children. The content of what students eat in schools has diversified and, according to some, deteriorated in recent decades as greater numbers of more sophisticated private food sellers have joined in the school feeding game. This has occurred both through the contracting out of regular food services (for example, according to Levine [2008, p. 181], the Marriott Corporation managed the lunch programs in more than 3500 public schools by the mid-1990s) and through the selling of food in competition with the USDA-regulated meal programs. The "ecology of policy" approach that frames this volume maps well these dynamics, providing a complex picture of what might best be called the politics of school food curriculum (see the Introduction).

Because of the diversity of programs, each funded through different combinations of public and private funding streams, each with its own system of accounting and reimbursement, and none with oversight from federal agencies, it is simply impossible to account with any precision for what is fed to inner-city schoolchildren in their school buildings before and after school. In my examination of such programs, the food provided varied dramatically, not only by state or district, but also from program to program and even from one local school site coordinator or parent governing council (where there are such entities) to another. During my fieldwork in Crossroads, for example, I visited 33 schools, and 10 of these schools I visited 3 to 8 times each over 6 months. Almost all of my visits took place during non-school hours. At various times I (along with parents and children in attendance) ate collard greens and fried chicken, baked potatoes and steamed vegetables, enchiladas and rice and beans, hot dogs and cupcakes, cookies and pizza, all sorts of salads and casseroles, barbecue and processed meat sandwiches, school lunch-surplus milk and juice, soda and fruit punch, fruit gelatins and potato chips, cakes and pastries, and—yes—candy. The food came from school cafeterias, from local small grocery stores, from mega-marts, from pizza companies, and from donut shops. Some of it was donated by local businesses, some was purchased by program staff, and some was brought by parents who were sometimes reimbursed and sometimes not. While there is much that is unknown—given the paucity of systematic research into this issue—about how school food in community-based, urban, contracted programs compares overall to community-based feeding in other moments, or how it compares to what kids eat during the school day, I can attest that what these programs serve is largely unregulated, varying widely based on highly local demands and circumstances.

How Are Children Fed?

This may seem like a silly question. We have all seen school cafeteria workers doling out portions of food to children; yet the way children are fed has changed a great deal over the last century. I focus here on what might be thought of as the "pedagogy" of feeding, which, just as the pedagogy in the classroom revolves around a relationship between teachers and students, involves at its core the relationship between feeder and fed. This section provides a cursory map of shifts in the pedagogy of school feeding over time.

Charitable feeding, which ran school feeding before the advent of universal school food programs, generally maintained a strict distance between those receiving charity and those providing it. Institutional feeding separated mealtime from the intimacy of family-based eating. The National School Lunch Program emerged from New Deal entitlement programs and carried a stigma with it in the prosperous years after World War II. There were people who were paying for services and others who were unable to pay but were sometimes provided with food nonetheless, and initially the decisions about who to feed and how to feed them were decidedly local. The attitude of the institutional food provider went from paternalistic to contemptuous in relatively short order. The state's failure to provide sufficient food for all children who were hungry was not considered a failure in the decades after World War II when American capitalism was expanding and the public image of the rising new (white) middle class was widely accepted. The federal government in fact ignored the expansion of hunger in the 1950s. And those who needed state help were often deemed freeloaders, failures in a meritocratic system. The racial tensions that whipped through urban schools in the wake of *Brown v. Board of Education* in 1954 added the element of racism to this brand of paternalism and stigma. Although the National School Lunch Program expanded through the Child Nutrition Act of 1966 to provide funding to feed all poor children, eligibility determinations remained a local affair for several years after this due to resistance to relinquishing local control. Poor children were fed more or less in accordance with the law, although, as Levine (2008) summarizes:

> Local decisions about which children received free lunches revealed a continuing legacy of paternalistic, if not racist, practices...eligibility for free meals was determined by school officials—teachers, social workers, or principals—or by other individuals in the community, including PTA [Parent Teacher Association] members, welfare case workers, or local ministers. (p. 118)

With local officials parsing the "deserving" and "undeserving" poor (Katz, 1989), the power-laden relationship between feeder and fed was one of contempt on one hand and embarrassment on the other. This continued to be the case until local officials ceded control in the late 1960s.

When the Black Panthers fed schoolchildren breakfasts in urban churches and other locations outside schools in the late '60s and early '70s, the relationship between feeder and fed was quite different. Black Panthers fed children as a way of taking care of the community, and there was pride and identity associated with providing for the community's survival. In fact, many of the Panthers ate with the children, using the act of feeding children and eating with their community as a way of grounding their political work in actual relationships with people for whom they were fighting (Churchill, 2001).

When school-based feeding became a universal, categorical entitlement in the late 1960s, the relationship between feeder and fed changed. Feeding was still institutionalized, of course, but there is reason to think that the contempt and guilt associated with school feeding of poor children in the 1950s and '60s shifted considerably as a result of several factors. These factors included a shifted ideology of urban development, the U.S. government's embarrassment by the international exposure of a great, largely black, urban underclass in its midst, and the cumulative efforts of anti-hunger activists in the 1960s. The federal government was, in the wake of the Civil Rights Movement, newly inspired to illustrate that as the great capitalist superpower America was well able to take care of its own. Feeding all hungry children became a government obligation, a matter of national duty, and a true categorical entitlement. Eligibility would henceforth be determined by federal economic poverty formulas, not by the will or mercy of local officials.

In contemporary before- and after-school programs that are administered under government contracts to private and non-profit organizations, what is the pedagogy of feeding? What is the relationship between feeder and fed? My research within these programs suggests that the relationship is considerably less institutionalized and standardized than school lunch programs. Some children are fed in classrooms or in small groups in a way that is integrated with structured activities, while others eat at an appointed snack time. While breakfast is often administered by cafeteria staff, often both before- and after-school feeding is administered much less formally by teachers or other staff, and frequently not by dedicated food services staff.

At the same time, feeding in contracted programs is often, though not always, a business affair. Hungry children are fed not solely as a moral obligation, a political conviction, or a requirement of law. Instead, hungry children are fed as part of the fulfillment of a contract. There was, in my observations, no stigma associated with needing food within urban before- and after-school programs. There are structural and logistical reasons for this lack of stigma. Before- and after-school programs need to serve needy kids in order to stay in business, and, because of the hours of service, they need to feed kids in order to serve them. Indeed, if it were not for low-income children and their many needs, to put it crudely, there would be no contracts and therefore no jobs for those who are currently paid to feed, tutor, and care for these children outside of school hours. In fact, there is a manifest desire and effort on the part of every publicly funded after-school program I have encountered to keep children and their parents satisfied. In part, this desire has to do with the need to maintain high attendance, for contract dollars are often based on numbers of children served per day. Thus, feeding is just one of many requirements of a contract and ways of attracting and maintaining "clients" for a service.

How children are fed and the relationship between feeder and fed—the "pedagogy" of feeding—in contemporary contracted services is thus a mixed story. Students in contracted school-based programs are fed in a way that is generally more intimate and without stigma than institutional food programs, similar in some ways to how the Black Panthers fed children. But they are also fed—as they are served more generally—in a way that is more akin to a business relationship than institutional entitlement feeding. For better or worse, in contemporary community-based school feeding programs, children are clients of the food providers. This tension between providers and their parent and child "clients" was obvious in the Coalition for Local Initiatives site council meeting. Parents needed the services, and they rarely had the collective power to determine who provided them. And parents consistently struggled with the fact that structurally, even if—like in the Coalition—there was a robust effort to build parent governing mechanisms, after-school programs are rarely accountable to the parents they serve. I have seen a few cases of truly and robustly parent-governed sites, but in the grand scheme of urban after-school programs these are decidedly the exceptions to the rule. The rule is far more akin to a business relationship.

For What Purpose?

Why do different entities feed hungry urban children in the United States? This is a fundamental question that is, surprisingly, rarely asked. Perhaps the answer is taken for granted—they feed hungry children because it is the right thing to do, or because children cannot learn when they are hungry. But motivation is a complex business, and different feeders have been motivated in different ways to address the hunger of children in the inner cities of the United States.

Consider the examples I have cited already. The original motivation for the institution of the National School Lunch Program was multiple, involving a desire to address hunger in the wake of the Depression, a desire to provide welfare for American agriculture, and myriad less savory political motives (Levine, 2008). The National School Lunch Program and the National School Breakfast Program also expanded in part out of a desire to address the national embarrassment of dire poverty and hunger. Second, the Black Panthers fed children in order to (1) keep party membership grounded in the community and its needs, (2) develop and maintain the good will of the community, and (3) demonstrate the systematic intention of the United States government to maintain the black community as a permanent underclass. Third, of course, contemporary contracted programs feed children partly because it is in their contracts to do so. Such programs' least savory motivation for feeding is, as previously mentioned, the self-interest to keep attendance up so as to maintain the size of the program and the contract dollars that go with it. Yet such an overall structural analysis misses important qualities about programs' motivations for feeding. Because these programs are so utterly decentralized, so irregular, they are also quite diverse. Different programs are comprised of different food decision makers, as discussed above. Thus, there are many different reasons for feeding hungry children within urban after-school programs, including of course the genuine desire by some actors to improve the lives of the children and community members served.

Some programs are governed in whole or part by parent volunteers, and to the limited extent that parents are in control, they are motivated to feed their children because, well, they are *their* children and, as is obvious in parents' dialogues in site council meetings like the one chronicled earlier, they care deeply about their children's well-being. As noted in the last section, however, most before- and after-school programs are controlled by professionals whose motivations for feeding vary. These professionals tend to talk about feeding in terms of its direct link with children's behavior, happiness, and performance during the hours children spend in the program. For example, they talk about children's attention spans, about warding off sugar rushes and crashes, about not wanting children to be listless in basketball or hyperactive in math tutoring. Occasionally, they also talk about children's hunger and nutrition as a good in and of itself, often noting that, for some of the kids, this afternoon snack will be their dinner as well. Administrators, in instances when they are the decision-makers regarding program feeding, talk about food in much more abstract ways. They discuss food as a logistical component of a program that has to be "dealt with" and "accounted for." Generally, the farther away from children people are in these programs, the more abstract is their feeding motivation.

Readers might wonder, if children are fed, what does it matter what feeders are thinking about? In diverse programs like those that characterize today's before- and after-school urban landscape, there is a great deal of flexibility in who has decision making responsibility for feeding children. This flexibility is both the opportunity and the danger inherent in decentralized, contracted urban school programming. Flexibility in contracts and lack of regulation and oversight implies both the ability to listen more to parents and the ability to listen less. This decentralized services model implies a tolerance for programs closely tailored to local contexts as well as a tolerance for programs standardized in accordance with the efficiency studies of a corporation. It means, for example, that several parents and a local non-profit organization might start a community garden at an urban after-school program with a dynamic and responsive program director. And it means as well that a major corporate junk food company might donate a year's supply of packaged, preservative-laden artificial foodstuffs to a penny-pinching program administrator who, highly influenced by dominant discourses of ignorant and uncaring parents, could not care less what parents think of the snacks. The reason for feeding, in other words, affects how and what is fed, along with whether those who feed are accountable to the parents whose children they feed. When it comes to much current before- and after-school feeding, these programs generally aren't accountable to anyone else.

Discussion

School food politics can be discussed as a purely normative concern. On the one hand, it is generally uncontroversial in the United States to proclaim that children should be fed more healthful food, that no child should go hungry, and even that school feeding should be integrated into the curriculum or that children should learn about food. From such convictions we can proceed to examine why this is *not* the case in most urban schools across the United States and proceed to address these many barriers. Indeed, this is what activists from school-garden advocates to Jamie Oliver are doing. But what are the relationships between who is involved in urban school food decision-making, how feeding takes place, who is fed, why they are fed, and what they are fed? What does this kind of critical analysis tell us about the politics of school feeding in urban U.S. schools? Here we might look at how the parent site council meeting discussed in this chapter contrasts with other urban after-school programs in which such parent meetings do not take place.

I have had contact with after-school and other contracted service providers in dozens of schools and neighborhoods in five urban centers over the past fifteen years, as a community-based arts program director, as a research assistant and volunteer, as an independent researcher, and as a professor and supervisor of student field experiences. Throughout these years, I have seen that sometimes parents' voices are regularly central to organizational decision-making processes, as in the case of the Coalition for Local Initiatives, which includes community organizing support and a sincere institutional effort to create space for ongoing parent involvement. Other after-school programs I have worked with—most, in fact—involve no formal parental role at all, so parents' negotiations of these programs are held in private spaces and actions. For example, I have seen communities of Muslim immigrant parents refuse to send their children to free, government-funded urban after-school programs that did not systematically incorporate parents' voices. These parents cited the ignorance of Halal food restrictions as one example of how their wishes were not taken into account. Some contracted programs incorporate the rich resources of their diverse urban families—including their foods—into the curriculum or program. Far more often, parent involvement in after-school or other contracted school programming, when it happens at all, is limited to parents' being pressured by their children to "participate" by purchasing candy for a program fundraiser.

Food is an act and a symbol of care, so concerns about food often become incorporated into parents' relationships with their children's after-school providers. I have heard many individual parents of children in urban after-school programs express hushed reservations about the content and staff of these programs, feeling isolated and alone in their concerns and having no formal means of involvement. I have seen similar parents, galvanized by a skilled organizer among them, gather together in a critical mass to exert pressure on a program, and even break off to start programs of their own in order to gain control and voice in their children's after-school activities. The meeting detailed above is but one example of the type of negotiation that happens when low-income parents are included in the process of meeting their children's needs in the much-expanded school days that characterize many inner-city children's routines. When parents are not included, their negotiations of the contracted programs that affect their children still take place, just in different forms.

In order to make claims for parental involvement, we must also be asking broader questions. Ultimately, who are the stewards of a new public vision of feeding in the contemporary urban landscape? Many people are profiting from school reform in every area, and feeding is no exception. Some private organizations seem intent on inviting collective ownership and responsibility for feeding reforms, pushing public schools toward a new form of public engagement. School gardens, educational school kitchens, and the involvement of diverse parents in contributing to an eating-learning curriculum around food suggest just this type of activism. Like the Black Panther Party, they advocate for public institutions to be accountable to the public—to all of our cities' diverse families, to all of our children—and they say that they will not sit by and wait for this to happen but will instead demonstrate what it looks like for communities to take care of their own children.

To some, these types of community programs can look inefficient, and for this reason I am concerned for the viability and scalability of such reforms in the current political, economic, and ideological climate. Under the great

American value of efficiency and quality control, so many proposed school feeding reforms simply blame individuals for failing to attain a certain degree of health. In their most malicious form, these reforms call for individuals to be held accountable to public and private institutions for the increased risk they present. We are moving through a political environment in which the discussion is, for many, not about how to make public institutions more accountable to the public, but rather about how to make each individual member of society accountable to the institutions and corporations that may eventually have to respond to widespread ill health. How will this ideological shift affect school feeding? How will it affect the programs that operate in conditions of (deepening) urban poverty?

Implications

The practical warning of this chapter is to pay attention to the details of reform and to be aware of how reform agendas to respond to conditions of urban poverty shape our ideas about the urban poor. Most contemporary education reformers begin by declaring their mission to improve the dire state of inequality or inadequacy in schools. They demonstrate the urgency of this dire state before presenting details of reform plans, selling the crisis rather than the solution. In just this way, school food reformers declare—usually dramatically—a desire to improve children's health, citing details of increasingly poor health as justification for immediate action along whatever lines they recommend. As with school reform generally, the details of the school feeding reform plan—not the poignancy of the would-be reformer's narrative of obese or malnourished children—are the only things that matter. Who feeds whom what, how, and for what purpose? We should be wary of any school food reformer who dismisses these questions as irrelevant, wary of those who emphasize that, as long as the reform "works" to efficiently produce healthy kids, we don't need to look at who is feeding, their motivations for feeding, what is being fed, or how the feeding system works. Like school reform, school food reform is simply not a "common sense" affair.

In the United States today, businesses profit from feeding poor children in the lunchroom, and contracted educational service organizations feed poor children in part to stay in business. In the United States, many public school systems have fed poor children out of charitable contempt, out of ideological conviction, out of indifferent obligation, as well as out of noble intention. Activist community organizations in the United States have fed poor children as a means of demonstration, out of a sense of ethnic pride and community solidarity, and out of a will to survive with dignity when the implicit answer to whether "we" will feed poor children—particularly poor children of color—was "no." And, in the contemporary United States, the legacy of these progressive, paternalistic, activist, profit-driven, and political-institutional dynamics can be seen in events like the meeting at the site council profiled here, where a more or less democratically run non-profit organization works to produce spaces where the content and method of feeding—among many other things—are negotiated. This space for negotiation that the Coalition for Local Initiatives site councils provide for the discussion of the details of feeding is, in part, a recognition that feeding is a powerful act. Not simply what is fed matters, but *who feeds whom what, how, and for what purpose*. The implication of this chapter is that we should consider all questions of school food reform in this broader context and, at the same time, that these questions of reform should be grounded in the specific contexts of actual urban schools and communities, within which low-income parents struggle for recognition, self-determination, and the health and well-being of their children.

References

Churchill, W. (2001). To disrupt, discredit, and destroy: The FBI's secret war against the Black Panther Party. In K. Cleaver & G. Katsaificas (Eds.), *Liberation, imagination, and the black panther party (pp. 78–117)*. New York: Routledge.

Demas, A. (2000). *Hot lunch: A history of the school lunch program*. Trumansburg, NY: Food Studies Institute, Inc.

Katz, M. (1989). *The undeserving poor: From the war on poverty to the war on welfare*. New York: Pantheon.

Levine, S. (2008). *School lunch politics: The surprising history of America's favorite welfare program*. Princeton, NJ: Princeton University.

Marine, G. (1969). *The Black Panthers*. Chicago, IL: New American Library.

Poppendieck, J. (2010). *Free for all: Fixing school food in America*. Berkeley: University of California Press.

School Nutrition Association. (2009). *School nutrition operations report: The state of school nutrition 2009*. National Harbor, MD: Author.

United States Congressional Committee on Agriculture, Nutrition, and Forestry. (2010). *Beyond federal school meal programs: Reforming nutrition for kids in schools*. (Senate Hearing 111–242). Washington, DC: U.S. Government Printing Office.

• CHAPTER TWO •

Fixing Up Lunch Ladies, Dinner Ladies, and Canteen Managers

Cases of School Food Reform in England, the United States, and Australia

Marcus B. Weaver-Hightower

Many people are out to change the way canteen managers in Australia, "dinner ladies" in England, and "lunch ladies" in the United States do their jobs and the products they serve. As obesity crises, particularly (e.g., Popkin, 2009), have gripped the attention of these Anglophone countries, school food—often seen as a major contributor to such crises (e.g., Mission: Readiness, 2010)—has been the subject of wide-ranging reform efforts. Whether slow food or grab-and-go food, organic food or fortified food, local foods or global cuisine, progressive attempts at expanding programs to three meals a day or conservative attempts to restrict or eliminate the programs entirely, varied reform efforts are competing to influence how governments and schools feed children.

In this chapter, I explore the context-dependent work of organizations advocating for school food reform in these three countries. From England I focus on the School Food Trust, the quasi-governmental organization charged with implementing the transformations wrought by high-profile school meals reform. From the United States I present legislative efforts by the Physicians Committee for Responsible Medicine to insert vegetarian and vegan concerns within legislative processes. From Australia, finally, I present the work of the Healthy Kids Association, a New South Wales state organization with significant influence around the country. The important similarities and key differences of each country and each organization are telling of the complicated politics surrounding school food. Synthesizing what might be learned from these cases can inform other reform efforts, ultimately illuminating key concepts about food, cultures, policy ecologies (Weaver-Hightower, 2008), and the educational project of these three nations.

I begin by outlining the methods used to construct the case studies analyzed. Then, I describe each case, surveying the national context as well as the mission and strategies of the three reform organizations. Finally, I present a cross-case synthesis, highlighting the complexity of reforming something so deceptively simple as school food.

Methods

The case studies presented here are developed from a large-scale study focusing on the politics of school food around the world. Methods of critical ethnography (Carspecken, 1996) were used. Research was conducted in South Africa (dropped here for space considerations), Australia, England, and the United States, all English-speaking countries and all currently focusing on reforming the food served in schools. Traditional qualitative procedures typified the data collection, including participant observation, interviewing, archival research, and document analysis. Each case varies slightly in the number of interviews, documents and observations obtained, but each has been rigorously checked for validity through triangulation of sources and, when feasible, member checking (Lincoln & Guba, 1985). I alone collected and analyzed the data, though I depended greatly on others' help in connecting with participants, finding documents, and transcribing interviews.[1]

The three cases presented here were chosen from among the many possible to illustrate—as Merriam (1998) suggests—the "hows" and "whys" of reform's complexity in differing countries. Each case was bounded by focusing on a single organization with a major mission to reform school food in a single country—Australia, England, or the United States. Each also has telling differences, particularly in the overall contexts in which they operate. To highlight this I have used the terminology and techniques of policy ecology analysis (Weaver-Hightower, 2008; see the Introduction, this volume). Finally, a cross-case synthesis (Yin, 2009) was performed to develop themes that illuminate the shared challenges of reform.

Case One: England: The School Food Trust

Overview of School Food in England

England's school food provision has the earliest emergence of the three countries. With the Education Act of 1870 making elementary education compulsory, poor children began attending schools in large numbers, and their health problems became a concern—a pressure on the system (Berger, 1990). The Education (Provision of Meals) Act of 1906 was the first organized governmental provision of school meals in England, though many voluntary agencies had already been feeding children. One of the main arguments for it was concern about the fitness of soldiers, given the major losses incurred by the British during the Boer Wars. The 1906 Act was not a mandate to provide meals, but it made it possible for local educational authorities (LEAs) to use local taxes for the purpose.

The postwar Education Act of 1944 dramatically changed schools in England, including restructuring LEAs, increasing the school leaving age and establishing a new secondary education system. It also changed school meal provision dramatically by mandating that LEAs provide a fixed price meal to every child who wanted one. Schools around the country had to expand or create meal services and many began to hire nutritionists and professional staffs to fill the needed niches and roles of an expanded service. A system of cafeterias also began springing up in both primary and secondary schools nationwide as a means of adaptation to these pressures. By 1947, the government paid 100% of school meals to those LEAs who met the conditions for poor children. The goal was still to provide meals free to all students universally. By 1952, however, the education ministry conceded that this was unlikely to ever happen given extant conditions, and prices for paying children began to climb thereafter.

Both Labour and Conservative governments of the late 1960s through the 1980s unraveled much of the school meals service. In 1968, Harold Wilson's Labour government withdrew free milk from secondary schools, and the Education (Milk) Act 1971, championed by then Secretary of Education Margaret ("Milk Snatcher") Thatcher, further withdrew free milk from students over age seven. (Like many other countries, in England provision for younger children tends to be more protected than those for older children and teens; see also Robert & Kovalskys, this volume.) The Education (No. 2) Act of 1980, Thatcher's major privatization reform as Prime Minister, did the most to dismantle the service and send it into entropy, mainly by removing the requirement that all schools serve meals. The Act left only two compulsory tasks for LEAs: (a) provide some kind of food for students with parents on welfare and (b) have a place for students to eat lunches brought from home. In the wake of the 1980 Act, kitchens closed, jobs were lost, and free meals took on special stigma, for often only poor children were given meals. The Social Security Act 1986 further converted the free meals system, adding the price of school dinners (lunches) to welfare checks rather than directly giving the meals, giving families a choice of whether to direct the money to school dinners; this resulted in hundreds of thousands of children not taking a daily meal. Further dismantling the service, the Local Government Act of 1988 instituted "compulsory competitive tendering," which forced LEAs to put the meals service up to the lowest bidder, the intention being to show the inefficiencies of public services compared to private industry. The quality of meals—especially given that nutritional requirements had been completely removed—declined precipitously as private caterers, who replaced unionized employees of local governments (*succession* in policy ecology terms), sought to maximize profit.

The school food service environment stayed much the same for years, but was thrown into tumult in February 2005, when a reality television program, *Jamie's School Dinners* (Gilbert, 2005), starring celebrity chef Jamie Oliver, aired on Channel 4. Though numerous organizations were already working to make school dinners healthier (see Morgan & Sonnino, 2008; School Meals Review Panel, 2005), it was Oliver's efforts to convince the dinner ladies in the Greenwich borough of London to stop selling turkey twizzlers and chips (french fries) that created a massive public outcry for improvements in school dinners (see Kang, this volume, for similar South Korean public outcry over food quality). In March 2005, Oliver was granted a meeting with Prime Minister Blair, who promised £15 million for the establishment of the School Food Trust, a quasi-governmental group charged with guiding changes in nutrition and other areas of school food service. In April 2005, the Labour government exceeded these expectations, promising £280 million (roughly US$440 million) of new inputs—£220 million for fresh ingredients and £60 million for the School Food Trust.

The School Food Trust

The School Food Trust is a "quango," a quasi-autonomous non-governmental organization, meaning a group to which the government has devolved some of its powers; it operates in concert with British government departments, with funding from the government, but it has its own mission and responsibilities. Despite its tenuous connection to government, the Trust became a target for political criticism, mainly because it was the public, organizational face for insisting on tough reforms in school food provision sparked by Jamie Oliver's exposé.

The School Food Trust's mission has been to facilitate the reforms called for by a number of governmental actions, including the School Meals Review Panel's (2005) report, which followed from the Jamie Oliver-inspired public concern. This role has spawned a multifaceted approach to increasing meal "take up" (the number of children consuming meals); addressing the "deskilling" of cafeteria staff; enabling compliance with food- and nutrition-based standards; conducting research to show the need for and efficacy of improving school meals; and communicating the standards, their research, and best practices to all stakeholders. This last task was no easy feat, for the actors that are their stakeholders are diverse and many, including the general public, schoolchildren, parents, teachers and administrators, catering managers and frontline dinner ladies, policymakers, LEAs, and food manufacturers.

The Trust created a raft of interventions aimed at making progress on these various niches and roles they filled. In working with caterers and cooks, the Trust provided cooking skills training through their School FEAST (Food Excellence and Skills Training) network. They also worked with caterers to develop menus that comply with new nutrient-based nutritional requirements and worked to support software development for this purpose, as well. The Trust website (http://www.schoolfoodtrust.org.uk) provided numerous case studies of best practices from around England. The Million School Meals program was a main thrust in trying to increase take up of meals (the main metric used in both public and governmental debates). The program's resources "include everything from marketing support to recipes, posters and curriculum packs, plus guidance for encouraging pupils to stay on site at lunchtime, and creating your own packed lunch policy" (http://www.schoolfoodtrust.org.uk/schools/projects/million-meals-resources). The Trust also provided branded advertising (promotions were done with Disney's *High School Musical 3* and Disney-Pixar's *Ratatouille*, for example) to help cafeterias market their food service.

Even with all of these services devoted to them, food service workers were initially the group of actors most resistant to reform, according to School Food Trust staff with whom I spoke. Caterers and dinner ladies took a great deal of blame in the wake of the Jamie Oliver campaign, and meal take up plummeted after the show, causing many of these workers to fear for their livelihoods. Some of the workers with whom I spoke five years after Oliver's show were ultimately glad for the changes to the food, but "Jamie Oliver" was still a dirty word to many; the School Food Trust was initially resisted as a part of Oliver's mission. The Trust's relationships with dinner ladies and their professional organization, the Local Area Caterers Association, has become far more cooperative in recent years, however, after long years of assisting frontline workers in good faith.

Working with students to encourage healthier eating was a strong stream of the School Food Trust's work, as well. The Trust created curriculum materials, a cooking skills program (Let's Get Cooking!), and online resources like games and informational brochures for children and their parents. They also sponsored programs to send chefs into U.K. schools and introduced competitive charitable programs to pledge money for school meals in high-hunger nations in other regions of the world.

Research was another major area the Trust focused on, using it as a means of backing up claims made about the importance of school food to the academic and social missions of schools. As the director of the research division told me,

> And if you know the background to the area [of research] then none of the [previous] data were very convincing, or what there is convincing really comes from developing countries and almost nothing from either the U.S. or Australia or Canada where you could say there's really strong, clear evidence that if children are eating better at school, for example, they'll do better on their exams or they'll behave better in the classroom. So our job was to sort of take forward that very broad agenda.

One of the more rigorous studies to accomplish this "broad agenda" was a controlled experiment to evaluate on-task behavior for schools that implemented changes to their food and their dining environments (see School Food Trust, 2009b, 2009c). The Trust found, among other things, that students in schools that had made changes were three times as likely to be on-task as students in control schools. Other research tracked the take up of meals nationally, procurement trends, nutritional standard compliance and the attitudes of various stakeholders (for a summary, see http://www.schoolfoodtrust.org.uk/research/the-trusts-research).

The communications division of the Trust was responsible for monitoring and responding to the media coverage of school meals, a tall order given its frequency and level of scrutiny, particularly post-*Jamie's School Dinners*. The Trust was a frequent contributor of comments for journalists, and this gave it a high profile as avatar of the "nanny state" among conservatives, as a group that was trying to "preach" to parents about packed lunches and trying to "take away" beloved, culturally prized foods. Conservatives saw the School Food Trust as a high profile holdover from the Labour government that installed it, and the strong libertarian segments of the English public have long been distrustful of the food bans and regulations the Trust championed.

Even with detractors, the Trust has shown promising results over its existence. Take up of school meals increased, up 2.1 percentage points in 2010 from 2009 (Nelson et al., 2010). Not only are students eating, but they are eating healthier options, with the amounts of fruits and vegetables consumed going up since 2005 (School Food Trust, 2009a). As a high level director told me, there were other important changes, too, including a wider variety of foods for students and a sense that cafeteria staff are not as marginalized. Educators now, he said, "recognize that there's a contribution, that the...cooks and the dining room staff make to children's well-being in a number of different ways." He also noted an important change in the effort and attention being put into food and child health from educational leaders:

> We just did some telephoning around very briefly over the last couple of days to talk with…about twenty local authorities who had said that they thought they would be compliant with the standards. Five years ago if you'd rung up those same twenty local authorities, you'd have been lucky if you'd had a conversation with five of them, I would think, about the quality and provision of food in schools. (…) They wouldn't have been able to answer that question at all and I think now probably three-quarters of them can, and with some depth, and with feeling. So I think there's been a sort of sea change in the ways in which not only schools, but also local authority caterers, are engaged in this, and the private caterers as well.

Part of this new attention has sprung from successful research, described above, that has shown some promising connections between school food and academic and social success. The Trust's work has demonstrated that reform can have broad impact, an appealing fact for administrators in the competitive context of school choice in England.

In the two school cafeterias I visited in March 2010, the changes were visible. Very few frozen and processed products were used; instead, the kitchens made food from scratch, using fresh meat, fresh vegetables, and low fat and low sugar recipes. The dinner ladies—one of whom conscripted me as a cook for the day—peeled, chopped, seasoned, fried, and baked rather than simply opening boxes and reheating. Some of the food was very good (some, admittedly, was not), but regardless the food was not processed and it was fresh and many students and teachers ate it gladly, all for a mere £2 (about US$3) for a full price meal—a bargain in comparison to the rest of London. These are norms for most schools in England as of this writing, and the proof of positive change was in the pudding, quite literally.

Despite these very promising results, after the May 2010 elections brought the Conservative-Liberal Democrat Coalition to power, the Trust's position as the face of school food reform made it a prime target for predation. Within weeks, the Coalition cut the Trust's budget by more than £1 million. Plans that Labour and the Trust had made to extend free school meals to 500,000 more students were scrapped. The new Education Minister, Michael Gove, announced that academies, the Coalition's unregulated, community-run schools (somewhat like charter schools in the United States)—which they hoped would be the new model for thousands of schools—would not be required to meet the nutritional regulations. The coalition government also shifted the School Lunch Grant, an £80 million-per-year fund specifically set up to help schools meet the direct costs of school meals, to instead be a general grant that schools can decide to do anything they want with; given the constant financial crisis of schools, many will no longer focus these funds on meal services. Finally, in late September 2010 (as I first drafted this article), the School Food Trust was put on a list of quangos to be abolished by the new government. As it is also a charity, the Trust can continue by raising its own funds, which it plans to do, but with few government connections and the threat of no regulations left to enforce, this seems a difficult conversion at best; to do this, they plan to begin offering services for fees (School Food Trust, 2010). Five years and hundreds of millions of pounds spent on reform to school food has been seemingly unraveled in about six months.

Case Two: The United States: The Physicians Committee for Responsible Medicine

Overview of School Food in the United States

England and the United States have very similar histories of school meals provision. Early meals were similarly provided by social progress groups throughout the late nineteenth and early twentieth centuries (see Levine, 2008). Finally, after decades of advocacy, these groups were able to convince the U.S. government to get involved in and increase inputs for school food. Like England, this occurred during the postwar Keynesian welfare revolution, specifically with the United States' passage of the Richard Russell National School Lunch Act in 1946.

Also like England, concern for malnourished youth unfit to be soldiers was a mainspring for federal action (Levine, 2008). President Franklin Roosevelt's National Nutrition Conference for Defense in 1941, for example, discussed the issue, and it led to the first federal nutritional guidelines. Indeed, the National School Lunch Act itself describes military readiness as a main purpose:

> It is hereby declared to be the policy of Congress, *as a measure of national security*, to safeguard the health and well-being of the Nation's children and to encourage the domestic consumption of nutritious agricultural commodities and other food, by ... providing an adequate supply of food and other facilities for the establishment, maintenance, operation and expansion of nonprofit school lunch programs. (§2; emphasis added)

The ordering of these priorities was no accident, for, at the time, the threat to peace from unfit soldiers was freshly in mind.

Creating and passing the National School Lunch Act was not easy, and its implementation has been similarly difficult. Political fights erupted over its states' rights implications, with southern states concerned that the federal government was overreaching and would use school meals as a way to interrupt segregation traditions and laws. Even after the law was passed, the spread of free school meals was slow, and there were definite racial and social class patterns to who was not getting fed. Further, the program had been put under the control of the U.S. Department of Agriculture (USDA), and it has been mired in agriculture politics ever since. The USDA has supplemented tight school food service budgets by installing a surplus commodities program that provides, according to the USDA, 15 to 20% of the food given to students, but this leaves the National School Lunch Program (NSLP) with ambiguous priorities; is it a program to help farmers or to help children, or can it do both? The late 1960s saw widespread public concern over hunger across the United States, and school food was a primary policy lever for fixing this malady (see Poppendieck, 2010). The National School Breakfast Program emerged in 1968 in response to this pressure, and conservative President Nixon presided over the largest expansion of that and the NSLP in their histories (see also Sandler, this volume).

Also just as in England, the 1980s saw repeated attempts to eviscerate the program (*predation* in ecology terms). President Reagan's USDA became infamous for trying to count ketchup as a serving of vegetables, mainly as a way to cut costs. While the worst of the cuts were avoided during the tenures of Reagan and Bush, Sr.—partly because of gaffes like "ketchup is a vegetable"—conservatives still pushed through the Omnibus Budget Reconciliation Act of 1981, which slashed subsidies for full price meals, raised price limits for reduced price meals, lowered eligibility for free and reduced meals, and made the eligibility process more difficult. This resulted in rather dramatic entropy, for 2700 schools dropped out of the program, more than one quarter of full price students dropped out, and many fewer students were eating reduced and free meals. Perhaps most damagingly, though, the NSLP has increasingly come to be seen as a welfare program after the 1980s discursive attacks on the program (Poppendieck, 2010, p. 73). The U.S. public's antipathy for taxes, too, has since kept the program on razor margins—an extreme form of prolonged conservation.

Recent U.S. concerns with school lunches have grown out of a decline in the quality of food served because of the cost cutting measures required of school nutrition personnel. School cafeterias have on average one dollar per plate to buy ingredients after labor and other costs have been subtracted (School Nutrition Association, 2008), so highly processed food is prevalent. Fast food brands and fast food fare are also widespread (School Nutrition Association, 2009). Numerous groups have thus taken up food reform efforts (e.g., Cooper & Holmes, 2006; Kalafa, 2007; Waters, 2008), and Jamie Oliver also came to the United States to reiterate the lessons of his previous reality series, this time in his Emmy award-winning *Jamie Oliver's Food Revolution* (Seacrest & Smith, 2010). This has provided public attention that groups, like the Physicians Committee for Responsible Medicine, have used to further their agendas for school food.

The Physicians Committee for Responsible Medicine

The Physicians Committee for Responsible Medicine (PCRM) was founded in 1985 as a group of physicians and laypersons committed to ending the use of animals in scientific research and to promoting the health benefits of vegan and vegetarian diets. The overarching concern for animal rights and diets that reflect that concern has led the PCRM to focus significant resources and policy attention on school food (see also DeLeon, this volume). While not as big as some other organizations involved in school food in the United States, like the School Nutrition Association (SNA), the PCRM has nevertheless attracted much attention for its efforts, largely due to high profile endorsements of its initiatives (detailed below). PCRM provides an interesting contrast to the other two cases in this chapter because it is a group wholly outside the government, is dependent on donations, and is trying to accomplish legislative change.

PCRM has two major strands of its work focused on school meals, both of which fall under the umbrella of its Healthy School Lunches campaign (http://www.healthyschoollunches.org). The first major strand is outreach to child nutrition professionals across the United States. The aim is to increase the number of "plant-based" options in schools, meaning vegetarian and vegan entrées and alternatives to milk. As the nutritionist associated with the program told me, "what we do is just help schools navigate the...menu planning system by the USDA and situate that with the vegetarian and vegan needs." Largely this involves visiting nutrition directors—who would normally contact them—consulting with them on products, recipes, and other practices that can help meet the needs of vegan and vegetarian students and reduce dependence on animal products. For many food service directors, the nutritionist said, the concern is meeting USDA nutritional guidelines, especially for protein, keeping things within cost, and making sure that children will eat the food. Much of PCRM's work is showing directors how they can meet these challenges.

> This year we've worked with school districts and gotten them to put a vegetarian item [on the menu] and the sales have been great and the kids have been receptive. So it's

> more just telling them what's out there, reassuring them that it'll be fine and the kids will enjoy it and it'll really be a good thing overall.

Another challenge for her is reassuring directors that

> it's not just a vegetarian that will benefit from this. Any student, no matter what their eating pattern, is going to benefit from having a meal that is lower in fat, higher in fiber, and full of fruits and vegetables and whole grains. It really is a healthy eating pattern overall.

There is a pervasive sense when talking with PCRM staff that they do not want to be pigeonholed as a fringe group. Their approach is to focus on universal health benefits rather than moral imperatives of animal liberation, and they focus on eating more plants rather than eating no meat or cheese.

A healthy portion of their work with child nutrition professionals is also outreach, targeting those who would not necessarily know to come to PCRM for consultation. These interventions include, for example, developing curriculum, marketing materials, and sample menus and bulk recipes that nutrition directors can use. One of their biggest outreach events each year is a dinner at the SNA's Annual National Convention. I attended in 2009, and it was a gala affair. A gourmet vegan dinner was served in a ballroom, and they had advertised the event in advance for all the thousands who registered for the convention. During the lavish three-course meal—which was clearly meant to show that vegan didn't mean boring, tasteless food—the PCRM's director and the nutritionist noted above gave research-based presentations meant to convince attendees of the benefits of a plant-based diet, the drawbacks of a meat and cheese-intensive diet, and some tips for how to incorporate more vegan and vegetarian foods that kids will eat but that will not break the budget.

The PCRM, much like the School Food Trust, also conducts research to help support its messages. It publishes School Lunch Report Cards (http://www.healthyschoollunches.org/reports/index.cfm) that score sampled districts on meeting or exceeding USDA guidelines, obesity and disease prevention efforts, nutrition education, and the extent to which vegan and vegetarian options are present daily. They also present Golden Carrot Awards annually to food service professionals for similar accomplishments, thereby building case studies of successful practice that other districts might follow.

The second major strand of the PCRM's work on school lunches is legislative activism. In the most recent five-year renewal cycle for the Child Nutrition Act—one of the major pieces of legislation, along with the Farm Bill (see, e.g., Imhoff, 2007), that lays out the rules of and funding for school nutrition—PCRM developed and advocated for an amendment called the Healthy School Meals Act of 2010 (H.R. 4870; see http://www.healthyschoollunches.

org/legislative/hsma.cfm). They got Colorado Representative Jared Polis, a vegetarian himself, to sponsor the bill, and they worked to secure more co-sponsors; as of June 2010, 65 representatives had co-sponsored.

As the PCRM's policy manager spearheading the effort told me, there are four main components to the Act. The first calls for a USDA "pilot project" to "research and develop plant-based options—plant-based protein products—that will pretty much take center of the plate entrée space, instead of your meat-based option." The research would focus on discovering the products that are "easiest to use, easiest to prepare by the schools, and with all the nutritional components that they're looking for. So basically research finding out what schools will use and like, and what kids will like." As she told me, focusing on such research projects usually proves less controversial than funding changes before research is done; they got the idea from those who advocated for more whole grains in the NSLP.

For the second component of the Act, those plant-based products found to be effective in the research would be added to the commodities list,

> and that's important because, commodities...are offered at a cheaper rate to the schools than the regular food that they purchase. So by putting it on the commodities list, it's more affordable and easy to identify by the schools.

Being on the commodities list would make it possible to find meat alternatives and to provide them without the relatively high out-of-pocket costs they now incur for tight budgets.

The third component of the Act adds a further incentive for schools to offer a vegetarian option every day: a 25% credit to each participating school's USDA commodity dollars. If 20% of all schools participated, this incentive could cost $46.6 million annually. To contextualize, 63.9% of districts offer a vegetarian option "on a consistent basis"—not necessarily every day—at "any" of their schools now, and only 20.5% offer a vegan option (School Nutrition Association, 2009).

Costs like this incentive are a political concern amidst the extant conditions of economic recession, but the most controversial aspect of the Act is the fourth, the mandate to provide milk alternatives (like water, juice, or soy drinks) and to make meals served with these reimbursable. Currently, a meal is considered complete and thus schools are paid back for them only if cow's milk is provided, and schools are required to provide alternatives only if students provide a doctor's note for allergies or other health concerns. Vegan meals, in other words, lose money for the school—a powerful disincentive. Much of the opposition from legislators PCRM has met with has been to this milk issue. As the policy director explained,

> There has been some resistance, perhaps, from—I wouldn't say 'resistance' but 'reluctance'—to sign on from people in dairy states [laughing] because they think their constituents won't support it or it might, you know, encroach upon their market somehow, then they don't want to support something. For example, Wisconsin, there's a lot of cheese. Things like that.

PCRM, she says, doesn't see reimbursing milk alternatives as taking away market share; "we see it as providing an additional option."

Much of the traction the PCRM had gotten on the bill, in getting it included in the House of Representatives-passed version of the Child Nutrition Act and getting a great deal of grassroots action through petitions and letters to representatives, came from some very high profile advertising and endorsements. One of the first was a Washington, DC, subway ad that featured a young African-American girl asking, "President Obama's daughters get healthy school lunches. Why don't I?" (see Kilpatrick & McCann, 2009). The ad got national attention in the United States because presidential children are often considered "off limits" in political debate. Still, the ads netted many new volunteers to the PCRM's cause because of its visibility. PCRM also has gotten publicity from high profile celebrity endorsements, including actors Hugh Jackman, Scarlett Johanssen, and Tobey Maguire, along with Jillian Michaels of *The Biggest Loser* reality TV series, former professional basketball star John Salley, and comedian Sarah Silverman. Aided by these endorsements, PCRM has gotten about 125,000 signatures on their petition supporting the Healthy School Meals Act according to the policy manager, and they have gotten more than $8.5 million in contributions and donations (http://www.pcrm.org/magazine/gm10winter/2009fiscal.html).

Despite PCRM's successes in getting the Act passed in the House version of the Child Nutrition Act, the garnering of celebrity endorsements, and a great deal of grassroots financial and advocacy support, the fate of their Healthy School Meals Act is uncertain. It looks, at the time of this writing (the fall of 2010), that the much smaller Senate version of the reauthorization might be passed instead, meaning the PCRM's amendment will not be passed. It may be the victim of a tight legislative schedule, an unwillingness to extend budget deficits, and conservatives' antipathy toward programs, like the NSLP, that are perceived solely as welfare. The PCRM's nutritional policy manager had a realistic attitude, though:

> There's so many bills for school lunch improvement, basically—I don't know the price tag on all of them—but it's like impossible for all of them to make it in and get funded. So there's going to have to be a lot of decision, and it's going to come down to who has the loudest voice on it, and the other politics of how these things work.

I asked what she would do if the bill did in fact fail, and, rather than feeling defeated or down (though she will be disappointed), she said that they'll just try again later, maybe for the next Farm Bill.

Case Three: Australia: Healthy Kids

Overview of School Food in Australia

Unlike the other two nations discussed here, Australia's school food history is decidedly local, without significant federal involvement (a high level of *adaptive decentralization*) and without a program of free school meals for the poor. Rather, Australia's school food ecology centers on the school canteen or "tuck shop," often a small room with sparse cooking equipment that serves sandwiches, fruit, candy, drinks, and other quick-service, often highly processed foods to those children with cash to pay for them.

Indeed, Australia's school canteens have traditionally operated as profit-making ventures, set up and usually run by each individual school's parents and citizens group (P&C; like parents and teachers associations [PTA] in the United States) to fund school materials and activities. Though some canteens are now managed by private contractors, the large majority have highly local boundaries; they are supervised by a canteen manager (usually relatively low paid) hired by the individual school's P&C and staffed by parent volunteers, usually mothers. As women have increasingly moved into the paid workforce, though, frequent concern has been expressed in the media over the inability to find parent volunteers (e.g., *Sunday Telegraph*, June 24, 2010, "Schools lose food struggle"), and numerous canteens have moved toward entropy, having to close or restrict hours.

Their profit-making status means that the school canteen does not serve as a means of feeding a school's or community's poor children. While the meal services in England and the United States have historically had social welfare as a manifest mission, Australian school canteens have acted only as an aid to what is otherwise parents' responsibility for feeding their children at school. Huge numbers of school canteens are not even open every day; they may be open only on special occasions or just certain days each week, depending on what schedule is profitable. Poor and hungry children typically fend for themselves, though some schools give poor children money to spend at the canteen or provide a sandwich to children who came without lunch.

Health concerns and the inspection of school canteens has historically been a state-level rather than federal responsibility, just like most other educational issues in Australia. The federal government has only recently—since April 2008—begun a National Healthy School Canteens Project (http://www.

nhsc.com.au) to develop common nutritional and safety guidelines for school food. At the time of this writing, however, it looks as if these guidelines will not be compulsory. That means that the status quo of state-level organization responsibility will continue, with groups like Healthy Kids Association, the New South Wales school canteen group, leading reforms.

Healthy Kids Association

Like the School Food Trust and PCRM, Healthy Kids Association is largely a service organization, providing advice and consultation to schools that are interested in reforming their practices. Established in 1991 as the New South Wales School Canteen Association, Healthy Kids was formed in response to a perceived need among canteen managers for help in making canteens healthier. Since then Healthy Kids has grown into a nationally recognized organization focused on canteen reform. Like the School Food Trust in England, Healthy Kids is frequently called upon to comment in media stories about school food and it is a major voice in policy debates.

Healthy Kids has three major aspects to its work. The first is working directly with school canteen managers to help them meet the nutritional guidelines laid out for schools by the New South Wales state government in its Fresh Tastes @ School Strategy (New South Wales School Canteen Advisory Committee, 2006). The core of the state strategy is a "traffic light" food system—based on calories, fat, sodium, and fiber—that defines "red" foods that should be eaten only "occasionally," "amber" (or, yellow) foods that one should "select carefully," and "green" foods that should "fill the menu." "Red" foods can be served only twice a term, and "red" drinks, like soft drinks and energy drinks, have been banned (though, according to Healthy Kids staff, noncompliance with such bans is rampant). Healthy Kids helps the school canteens meet the guidelines through a once-a-term magazine and their website, but they also have school site visits that they perform for member schools. As the director described it,

> We'll go into a school and we'll do a complete review of their canteen, and provide them with a report and recommendations and a whole lot of other relevant operational and food information....And there are some schools that we work with that we might spend you know 80 or 100 hours in, actually in hands-on work.

This work involves advice on business practices—ensuring profitability—just as much as nutrition.

The second major aspect of Healthy Kids's work is providing services to food manufacturers. Healthy Kids's industry team works with manufacturers to help them meet the Fresh Tastes guidelines as well as Healthy Kids's even stricter nutritional standards, and they help manufacturers understand the needs of canteens. As a member of the industry team told me,

> what we do is look at the NIPs [nutrition information panels], taste test, labeling, and then work with the company and advise them about the barriers in canteens, like the price and the, you know, how usable they are and storable their product is.

These efforts center on the "amber" range of foods. This generally means moving products barely "red" into the "amber" range, for "green" ratings are hard to get and manufacturers of solidly "red" foods find little use in trying to move into the school canteen market anymore. In many ways, then, Healthy Kids helps manufacturers reformulate already highly processed foods, but costs and storage and preparation are important elements that Healthy Kids can shed light on from a canteen's perspective, too. This process of working with manufacturers, according to one Healthy Kids representative, has helped manufacturers move away from many of the bad practices of years ago.

There is a great deal of interface between these first two major aspects of Healthy Kids's work, services for the school canteen and manufacturers. One of their biggest projects is the publication of a *Buyer's Guide* each year. Manufacturers who are registered members can have their "green" and "amber" products listed in this guide, and schools across Australia—even though Healthy Kids is specifically a New South Wales organization—get a copy. This is an advertising opportunity for industry, of course, but it is also meant as an aid for busy canteen managers, for Healthy Kids has a certification program that ensures canteens of the "green" or "amber" status of the products without their having to read the labels (a service for which manufacturers must pay). Healthy Kids also runs an annual "food expo" trade show that serves the dual purpose of giving manufacturers an opportunity to promote their registered products and providing canteen managers a day of professional development. Healthy Kids also is a contractor for several canteens themselves, which they run as test cases, trying out products and practices that they can then share with members, whether manufacturers or canteen managers.

A third line of work for Healthy Kids is conducting nutrition education programs, focusing the organization on the public health mission of canteens and schools generally. One obvious example is their website, which has comprehensive resources for nutrition information and links to government reports, classroom resources, and marketing materials. They also administer a program called Crunch&Sip in which registered schools provide a snack break for fruits and salad vegetables along with water during the school day. They provide materials to support schools in implementing such programs. They also have worked with a group of "outback GPs"—rural doctors—to bring nutrition education and cooking knowledge to remote, often highly indigenous communities.

One thing Healthy Kids does not do explicitly is advocacy. They have instead, the director told me, a philosophy "to work collaboratively."

> There are many advocacy bodies who try to effect change through brickbat rather than "OK, where's the middle ground here? What might we do? How can we change what's happening here?" And so that's quite an interesting position to be in because it's about balancing credibility and professionalism in the face of organizations who think you're only credible if you're demonizing the food industry....Our philosophy is to try to work cooperatively with all.

Healthy Kids thus treads a fine line between politics and service expertise, trying to be conscious of the multiple stakeholders they are chartered to serve. They cannot be seen to be against public interests, government interests, industry interests, or school canteens' interests. This is a difficult task given the volatile political nature of school food, but they have been in many regards successful; they are highly regarded nationally and sought out by the government for advice and by the media for commentary, perhaps much more so than any of their other state-level peers.

What The Cases Together Show Us

Each of the three countries outlined and each of the three organizations profiled—School Food Trust, Physicians Committee for Responsible Medicine, and Healthy Kids Association—are different in crucial ways. Each confronts a history and a sociopolitical reality that are unique from the others, and each has adapted strategies and tactics to fit their unique ecology. Still, there are also clear, common lessons that a synthesis of these cases can teach reformers everywhere. I turn to just some of these many possible lessons in this section.

Each Ecology Requires Its Own Unique Reforms

That organizations would have to adapt to satisfy their context might sound obvious, yet the ways in which this happens can provide insight into the complexities of reform. For instance, all three countries evince a strong ideology that food should be made appealing to students so that they will purchase it; in other words, people in these countries generally believe strongly in student choice in neoliberal, market-based terms. (Not all countries have this discourse as a guiding feature; in South Africa's township feeding scheme discussed in the Introduction, for example, the offerings are almost always a variation of corn-based gruel, and neither students nor parents expect to be given a choice.) Each reform effort above has had to negotiate this student choice discourse. The School Food Trust had to factor in student appeal because of the monolithic accountability requirement of "meal take up," and the Trust has thus worked on cafeteria environment and food freshness and appearance. PCRM has had to counter concerns that vegetarian food is unappealing and won't be cost effective, requiring that they fund research to establish what vegetarian foods are appealing so they can get those on the commodity list. Healthy Kids, finally, has had to consider profitability in an era of dwindling days the canteen is open and ease of preparation in a context of dwindling parent volunteers, and this has led them to work with manufacturers and schools alike to find foods that kids want to buy. Other issues demonstrate that ecologies influence tactics, too. Profit expectations, the relative deference to religious food cultures, tolerance for governments banning foods, relative concern for children in poverty, and many more issues hold tremendous sway over the approaches and strategies that reform organizations use.

Multiple Actors and Levels of Politics Must Be Navigated Simultaneously

From the federal to the local, school food reform efforts must navigate several different terrains simultaneously. Each organization's reforms has required the cooperation of school food service personnel, students, administrators, teachers, communities, government and industry. Each of these actors has differing interests, sometimes amenable to cooperation but sometimes competing. "Junk" food is a clear case in point. While governments feel pressures from communities in all these countries to restrict food of minimal nutritional value, kids love it and buy it. Their buying it is good for manufacturers and for those educators who get a percentage of the profits from it. Nutritionists and other public health officials, on the other hand, have good reason to worry about the mass profusion of such foods. Reform-oriented organizations are forced to deal with the politics of all of these levels. Healthy Kids, for instance, works with the state government (both the health and education ministries), local schools, federal panels, the media, and manufacturers, all at the same time, and they work carefully to balance these competing political environs.

All Ecologies Have Entrenched Interests and Traditions Resistant to Change

Food politics often involve deeply held and competing beliefs that must be negotiated carefully and with tact—beliefs about the role of government, about food, about poverty, race, and gender, and more. At an individual level, actors must be won to the causes of food reform. Children and adults alike resist campaigns to limit the foods they love but that aren't good for them. They also resist campaigns that seem to say, "Your food choices aren't good (or good enough)," whether in nutritional terms or in cultural ones. The reform organizations in this chapter have had to tread carefully with their arguments, trying not to offend key constituencies in these ways.

More than personal interests are involved, though; institutional and economic interests are also at stake. In highly developed Western countries, food manufacturing, for instance, is a massive industry. In the United States alone, the larger food system involves around 20% of the total workforce and about one trillion dollars (Belasco, 2008, p. 20), and, as discussed in the Introduction to this book, schools make up tens of billions of dollars of this market. Even with this amount of money available, there is more competition than ever for food dollars, and manufacturers thus fight hard to resist bans of their products (Nestle, 2007). Nutrition directors fight against unfunded mandates from government (one of the chief concerns PCRM heard about its Healthy Meals Act). Administrators, too, resist changes to vending or canteen offerings that reduce profits for their schools. All of these interests stand in the way of reform organizations. The School Food Trust, for instance, has had to work to overcome resistance from catering groups, the public, cafeteria staff, and the government to make progress on their mission.

Bans Are Out of Favor, so Organizations Avoid Them When Possible

Marion Nestle's (2007) *Food Politics* shows well what can happen when governments, even with solid health information to back them, urge people to "eat less." Agriculture and manufacturing groups—not to mention their governmental allies—resist such messages vociferously. This is perhaps especially true in the United States, where libertarianism and a deference to business interests is perhaps stronger than in the UK or Australia. This is why the PCRM has advocated for *increasing* plant-based options rather than *restricting* meat and milk.

In Australia and England, in contrast, each of the government-led reform efforts has included the restriction or banning of certain food items, particularly soft drinks. (Some school districts in the United States have done this, too, but this has not happened at the federal level.) Increasingly, though, the language of the School Food Trust and Healthy Kids materials mirrors the notions of "limiting" rather than banning. The traffic light system that Healthy Kids works with, for instance, does not prohibit red foods; it merely says they can be served only occasionally (twice per term—eight times a year). Loose enforcement, though—a clear sign of the reticence of governments to ban products—has kept many of the "banned" foods on the menu in dining halls and canteens in both countries. This was a constant source of work and frustration for both the School Food Trust and Healthy Kids.

Funding for Food Reform Is Often Tenuous

Most groups involved in school food reform, including those listed here, have tenuous funding situations. They must work hard to stay solvent. Being outside the government (or at least only loosely affiliated, sometimes working in symbiosis), school food reform groups are less able to guarantee their own funding and, thus, their long-term survival. Success, sensibility, and hard work, furthermore, does not ensure survival. This is true of both organizations, like the School Food Trust, and legislation efforts, like PCRM's Healthy Meals Act, where their efforts are entropying or coming to no fruition. For Healthy Kids, their future is currently more stable, but this may have more to do with being a membership-based organization that also provides paid services. They make their own money, and not just from the goodwill of like-minded members of the public (like PCRM must). Even Healthy Kids, though, must work to ensure it stays relevant to its members. As the School Food Trust moves toward paid services, they will face the same pressures.

Money Is Almost Always the Core Issue

Not only is the survival of reform organizations themselves dependent on funding, but the progress of particular reforms often center on money. The varied profit requirements of food service programs often require that they operate on razor thin margins, so any changes to the program must be proven cost-effective. Also, because of the massive numbers of children being fed, particularly in the United States and England, small improvements to meal services require substantial increases in public funding. As just one example, a mere six cents increase in per-meal reimbursement rates for U.S. schools in the recent reauthorization of the Child Nutrition Act (alongside some other minor changes) was going to require a $4.5 billion increase in allocations over the next five years. Making more substantial improvements to school food, seen in the context of recessions, constant economic crisis, and widespread antipathy toward taxes and government, is thus politically difficult. Those who would reform school food, then, must either scale back their ambitions or they must find the money (and other inputs) to realize their ideas.

There Is a Strong Focus on Assisting Frontline Staff

All the reform organizations in this study rightly focused on helping those who have to implement feeding programs, the cafeteria staff and food service directors. Though their relationships with frontline workers has not been without tensions (particularly for the School Food Trust), these are the people—because of the relative autonomy (or, *adaptive decentralization*) of local kitchens in deciding menus and procurement—who must be won over to school food reform. The PCRM works with child nutrition professionals in individual schools and districts to help them include plant-based options and to figure out ways to order them cost effectively. The School Food Trust trains cooks on how to prepare scratch food, a challenge given the deskilling wrought by the reheat-and-eat culture of the previous decades. And both the School Food Trust and Healthy Kids work directly with canteen managers and schools in meeting the government guidelines for nutrition. All of these professional development activities are vital, for without assistance many child nutrition providers would be capable of little more than the status quo; reform, on the other hand, requires new ways of thinking and acting.

Marketing Is Key

All of the reform organizations mentioned here have as one of their main focuses the promotion of their own agendas as well as the promotion of healthier eating in schools (however each defines that). Each focuses on those tasks quite well. For all, media exposure, whether traditional or online social media, was used as a tool to get their messages to their constituents and audiences. The PCRM, for example, got media exposure for their issues from their advertising and celebrity endorsements, but they also leveraged online technology well for fundraising. Hosting live events are clearly important, as well, for Healthy Kids runs a food expo to suit the needs of both canteen workers and manufacturers, PCRM puts on the dinner for attendees at the School Nutrition Association conference, and the School Food Trust has participated widely in cook trainings and conferences. Each organization also makes materials available on nutrition, ideas for cafeteria staff and directors, and materials to urge students to eat at the school (like the School Food Trust's *High School Musical 3* posters lauding the benefits of school dinners). Such coordination and saturation of messages keeps school food on the public radar and increases sales, a requisite for all the food service programs they work with. All three countries, moreover, are advertising-heavy environments, with particularly intensive marketing to children and teens; without marketing of its own, school meals could get drowned out in the din of commercial products.

Research Is Key

Each of the reform organizations covered here had as a challenge the need to communicate the importance of school food reform to the public, policymakers, some school administrators, food service professionals, and students. One means of argumentation they all chose was research, though of varying kinds. Healthy Kids began running their own canteens, partly as a means of testing out their reform ideas; this gave them more credibility, they felt, in working with other canteens because they could talk from experience rather than just "ivory tower" knowledge. PCRM, on the other hand, made USDA research on best practices a cornerstone of the Healthy School Meals Act, partly as a means of reassuring politicians and school nutrition directors that their plant-based options were feasible within the constraints of the program. The School Food Trust conducted its own research on links between school food and student performance, largely as a means of justifying the time, attention, and money being spent on reforms. Overall, each organization was seeking legitimation of their own ideas as a means of encouraging support and funding.

Working with Industry Is Complicated but Crucial

There is a strong sense among all three reform groups that no stakeholder is solely to blame for the state of school food and, following from that understanding, that no one sector can fix it or can be fixed so that all the problems will go away. Even the most arguably radical of the groups, PCRM, took a highly collaborative approach, wanting to work with those who disagree with them to see how they could overcome difficulties. They carefully crafted their approach, too, seeking out ways to make reform easy and profitable; for example, they worked closely with the USDA and food service directors on developing the Healthy School Meals Act, getting their ideas, advice, and best practices for how to make the act passable. More strident groups might have instead asked for regulation and mandates, no matter the impact on frontline staff. Overall, these three groups tended to look for the complexities of their cultures' food ecologies, and they believed that it would take work from all the various sectors—industry, government, food service, schools, and individuals—to turn around school feeding. Vilification of industry was generally eschewed, though industry did get its share of what criticism it deserved from these groups. PCRM, for example, urged manufacturers of vegetarian products to lower their prices—most of which are set artificially high for middle-class, health conscious shoppers—so that schools can better afford them. Healthy Kids and the School Food Trust also work with manufacturers on how to reformulate their products to meet nutrition reforms' pressures on schools. Many manufacturers have responded well, partly because it is in their financial interests to do so.

Conclusion

In sum, food reformers in the United States, Australia, and England exist in ecologies that are ever-changing but also tied to their histories. They are subject to multiple policies (health, agriculture, procurement, education, and more), but they often deal with ambiguous or nonexistent enforcement mechanisms. Each also works with multiple actors at multiple levels, and each is confronted with cultures that are at times amenable to change and at other times are fiercely resistant to healthier or more just eating practices. Because of these complexities they face every day, food reform organizations' survival and success are always in question. There is much that we can learn from them about food reform, including the practices they have implemented, the solutions they've proposed, and even the barriers and failures they've encountered along the way. Looking across these contexts in that way can give other reformers insight into new means of creating progressive food reform in their own contexts.

Note

1. Bekisizwe Ndimande, Becky Francis, Nicola Tilt, Jori Thordarson, and the staff at the National Library of Australia, the Library of Congress, the British Library, and the University of North Dakota have my undying thanks for helping with these tasks.

References

Belasco, W. (2008). *Food: The key concepts*. Oxford, England: Berg.

Berger, N. (1990). *The school meals service: From its beginnings to the present day*. Plymouth, England: Northcotte House.

Carspecken, P. F. (1996). *Critical ethnography in educational research*. New York: Routledge.

Center for Science in the Public Interest. (2007). *Sweet deals: School fundraising can be healthy and profitable*. Washington, DC Retrieved from http://www.csp_net.org/schoolfundraising.pdf

Cooper, A., & Holmes, L. M. (2006). *Lunch lessons: Changing the way we feed our children*. New York, NY: Collins.

Gilbert, G. (Director). (2005). *Jamie's school dinners* [Television series]. England: Freemantle Media.

Imhoff, D. (2007). *Food fight: The citizen's guide to a food and farm bill*. Healdsburg, CA: Watershed Media.

Kalafa, A. (Producer & Director). (2007). *Two angry moms* [Motion picture]. United States: A-RAY Productions.

Kilpatrick, K. & McCann, R. (2009, August 11). White House objects to poster that invokes Obama children. *The Washington Post*. Retrieved from http://www.washingtonpost.com/wp-dyn/content/article/2009/08/10/AR2009081003126.html

Levine, S. (2008). *School lunch politics: The surprising history of America's favorite welfare program*. Princeton, NJ: Princeton University Press.

Lincoln, Y. S., & Guba, E. G. (1985). *Naturalistic inquiry*. Beverly Hills, CA: Sage.

Maras, E. (2009, August). Recession drives profit protection initiatives. *Automatic Merchandiser, 51*, 28–42.

Merriam, S. B. (1998). *Qualitative research and case study applications in education*. San Francisco: Jossey-Bass.

Mission: Readiness. (2010). *Too fat to fight: Retired military leaders want junk food out of America's schools*. Washington, DC. Retrieved from http://cdn.missionreadiness.org/MR_Too_Fat_to_Fight-1.pdf

Morgan, K., & Sonnino, R. (2008). *The school food revolution: Public food and the challenge of sustainable development*. London: Earthscan.

Nelson, M., Nicholas, J., Wood, L., Lever, E., Simpson, L., & Baker, B. (2010). *Fifth annual survey of take-up of school lunches in England*. Sheffield, England: School Food Trust.

Nestle, M. (2007). *Food politics: How the food industry influences nutrition and health* (Revised and expanded ed.). Berkeley: University of California Press.

New South Wales School Canteen Advisory Committee. (2006). *Fresh tastes @ school NSW Healthy School Canteen Strategy canteen menu planning guide* (2nd ed.). Sydney, Australia: New South Wales Department of Health.

Popkin, B. (2009). *The world is fat: The fads, trends, policies, and products that are fattening the human race*. New York, NY: Avery.

Poppendieck, J. (2010). *Free for all: Fixing school food in America*. Berkeley: University of California Press.

School Food Trust. (2009a). *Research summary: Primary school food survey 2009*. Retrieved from http://www.schoolfoodtrust.org.uk/UploadDocs/Library/Documents/sft_primary_school_food_survey_2009.pdf

School Food Trust. (2009b). *School lunch and learning behaviour in primary schools: An intervention study*. Retrieved from http://www.schoolfoodtrust.org.uk/partners/reports/school-lunch-and-learning-behaviour-in-primary-schools-an-intervention-study

School Food Trust. (2009c). *School lunch and learning behaviour in secondary schools: An intervention study*. Retrieved from http://www.schoolfoodtrust.org.uk/partners/reports/school-lunch-and-learning-behaviour-in-secondary-schools-an-intervention-study

School Food Trust. (2010). *Information update November 2010*. London, England. Retrieved from http://www.schoolfoodtrust.org.uk/school-cooks-caterers/resources/sft-information-update-november-2010

School Meals Review Panel. (2005). *Turning the tables: Transforming school food*. London: Department for Education and Skills.

School Nutrition Association. (2008). *Little big fact book: The essential guide to school nutrition.* Alexandria, VA: School Nutrition Association.

School Nutrition Association. (2009). *School nutrition operations report: The state of school nutrition 2009.* National Harbor, MD.

Seacrest, R. (Producer) & Smith, B. (Director). (2010). *Jamie Oliver's food revolution* [Television series]. United States: American Broadcasting Corporation (ABC).

Waters, A. (2008). *Edible Schoolyard: A universal idea.* San Francisco: Chronicle Books.

Weaver-Hightower, M. B. (2008). An ecology metaphor for educational policy analysis: A call to complexity. *Educational Researcher, 37*(3), 153–167. doi: 10.3102/0013189X08318050

Yin, R. K. (2009). *Case study research: Design and methods* (4th ed.). Los Angeles, CA: Sage.

• CHAPTER THREE •

Cultivating Schools for Rural Development

Labor, Learning, and the Challenge of Food Sovereignty in Tanzania

Kristin D. Phillips
Daniel Roberts

> In the global era, is it wise to set, as policy goals, double standards for the rich world and the poor world, when we know that these are not different worlds but in fact the same one?
>
> –Paul Farmer & Nicole Gastineau (2009, p. 155)

Around the globe, policymakers and international donors have hailed school cultivation–the growing of plants and produce in the context of the school curriculum–as a cutting-edge approach to learning and charged it with confronting a variety of food-related social, political, and economic issues (Desmond, Grieshop, Subramaniam, 2004; FAO, 2004a). Yet despite school cultivation's status as a "cutting-edge" pedagogical strategy, school farms and gardens have existed in communities around the world nearly as long as common schools themselves. And despite the seeming international consensus that school cultivation can enhance learning and community development, approaches to school cultivation have diverged immensely across local and national socioeconomic contexts.

In this essay, we contextualize the current international aid vision of school cultivation within the history of one national context, the United Republic of Tanzania. We ask two questions: (1) What is the historical, social, and policy context for school cultivation initiatives in rural Tanzania? and (2) To what extent do the international policy vision for school cultivation for food security and current school cultivation practices address the key issues contributing to hunger and malnutrition in East Africa?

Specifically, we turn to the long and rich history of school farms in Tanzania to understand how the practice and discourse of school cultivation has been translated, interpreted, and implemented across the vast political and economic transitions that characterize Tanzania's last century. As anthropologist Dorothy Hodgson (2001) has written on the Tanzanian context:

> Development practitioners . . . often suffer from . . . "historical amnesia" . . . Although many planners rely on the past in the form of "baseline" surveys in order to evaluate the progress of their projects, few consider the actual history of development itself in the places in which they work. Intent on working for change in the future, they ignore the transformations that have occurred in the past, especially in terms of prior development projects. (p. 11)

To address this tendency toward historical amnesia in development initiatives, we draw on our ongoing interview, archival, and ethnographic research in village sites in three diverse districts—Monduli, Singida Rural, and Lindi Rural districts—to study school cultivation in Tanzania. We integrate findings from the policy and research literature; archival research conducted by Phillips on food security and educational development at the Tanzanian National Archives (TNA) between 2004 and 2006; and summer follow-up research by Phillips and Roberts in Lindi and Arusha regions in 2010.[1] The argument that emerges from the data collected is that school cultivation curricula should aim at education that does not simply target the knowledge and skill deficit of rural people, but that also addresses the political, social, and economic orders that construct and produce those deficits in the first place. We embed our analysis within the conceptual framework of "food sovereignty," a policy framework for rural development that, in addition to promoting technical assistance for food production, incorporates concern for the political and economic relations that govern access to food.

Food Security and Food Sovereignty: Key Conceptual Shifts in Rural Development

Currently 850 million people in the world suffer from hunger and malnutrition; of that number 815 million live in economically developing countries and 76% live in rural areas (FAO, 2004b). Hunger and malnutrition are not new items on international development agendas. Since the World Food Conference of 1974, concerns for "food security" have driven national and international agricultural and rural development agendas (United Nations, 1975). Despite this consistency in concern, considerable shifts have occurred in national and international approaches to eradicating hunger and malnutrition. To provide a brief history of development orientations toward hunger, we summarize shifts in the notion of "food security" between 1974 and 1994 and describe the subsequent reorientation toward a goal of "food sovereignty" by some nongovernmental organizations, civil society organizations, and social movements.

The major change in the history of thinking about food security has been a shift from the global and national level to that of the household and individual. The World Food Conference of 1974 centered on a concern that the world food system could no longer address the needs of the world population. The report from the conference defined food security as the "availability at all times of adequate world supplies of basic food-stuffs...to sustain a steady expansion of food consumption...and to offset fluctuations in production and prices" (quoted in Maxwell, 1996, p. 156). The conference led to efforts to create national self-sufficiency in food supplies and to create financing for countries to meet unexpected needs to import food (Maxwell, 1996).

It was Amartya Sen's (1981; 1999) economic theory of entitlements that fundamentally altered the development paradigm for hunger issues (Agriculture and Natural Resources Team, 2004). Through analyzing the Bengali famine of 1943, Sen's work shows that widespread hunger and famine can exist even in the presence of an adequate national and international food supply. The question, Sen (1981) argues, is one of entitlement—the "relations that govern possession and use" in a given society (p. 155). He writes:

> Viewed from the entitlement angle, there is nothing extraordinary in the market mechanism taking food away from famine-stricken areas to elsewhere. Market demands are not reflections of biological needs or psychological desires, but choices based on exchange entitlement relations. If one doesn't have much to exchange, one can't demand very much, and may thus lose out in competition with others whose needs may be a good deal less acute, but whose entitlements are stronger. (1981, p. 161)

This perspective has pushed international finance organizations and national governments to consider not only the macro-perspective of food supply, but also the micro-level of social, political, and economic relations that circumscribe access to food.

In 1996, at the World Food Summit, governments affirmed a commitment to halving the number of hungry people by 2015. Nevertheless, between 1995 and 2005, the number of chronically hungry people in developing countries increased from 800 million to 852 million—an increase of nearly 5 million per year (Windfuhr & Jonsen, 2005). Such statistics have led to a search for new approaches and new understandings of the causes of hunger, and they have led a number of local social movements, civil society organizations, and nongovernmental organizations to call for a new focus on "food sovereignty" rather than "food security." The food sovereignty approach adds renewed emphasis to Sen's call for attention to questions of access to productive resources, political power, and life possibilities in addition to technical support for food production. Specifically, the call for food sovereignty draws critical attention to the emphasis—originating from wealthy countries—on industrial agriculture, livestock production, and commercial fisheries, rather than a focus on the needs of the smallholder farmers, pastoralists, and fishers (who comprise at least half of hungry people) for secure access to productive resources (Windfuhr & Jonsen, 2005).

Today, interventions to eradicate hunger seek to address four areas of concern: (a) agriculture (increasing food production); (b) nutrition (increasing the nutritional quality of food consumed); (c) poverty reduction (increasing access to health care, education and other social services); and (d) democracy (increasing access to decision-making, information, and productive resources). "Food security" initiatives have generally addressed just two of these types of issues: agriculture and nutrition. The movement for "food sovereignty," however, generally expands this traditional focus to also mandate attention to issues of poverty and democracy.

For example, La Via Campesina, an international social movement of self-identified peasants, small- and medium-sized producers, landless, rural women, indigenous people, rural youth, and agricultural workers, composed a list of seven tenets that illustrate the food sovereignty approach's focus on poverty and democracy:

1. Food is a basic right;
2. Genuine agrarian reform must give landless and farming people ownership and control of the land they work and revise land rights to be free of discrimination;
3. Care and use of natural resources must be sustainable. Efforts to do so must be free of restrictive intellectual property rights and must rely on security of land tenure, healthy soils, and reduced use of agro-chemicals;
4. Food is first and foremost a nutritional source and only secondarily an item of trade. National agricultural policies must prioritize production for domestic consumption and self-sufficiency;
5. Food sovereignty is undermined by multinational corporations' control over agricultural policy and multilateral organizations;

6. Food must not be used as a physical or political weapon; and
7. Smallholder farmers, particularly rural women, should have direct input into the formulation of agricultural policy at all levels. (Via Campesina, 1996)

In this paper, we take seriously the assertion by Paul Farmer and Nicole Gastineau (2009) that "If the primary objective is to set things right, education is central to our task" (p. 162). We build on the principle of food sovereignty to propose that school cultivation initiatives be restructured by attention to the politics and economics of hunger. In particular, grounding a school cultivation curriculum in the food sovereignty approach can incorporate conversations about the local organization of land rights, the sustainable and equitable use of resources, the contradictions of conceiving food as a right versus as a commodity, the global regime of food consumption and production, and the resolution of food conflicts into the agricultural and nutritional education which existing school cultivation programs have effectively strengthened.

School Cultivation in Global Perspective

In the industrialized countries of regions like Europe and North America, school cultivation initiatives have generally taken the form of what is known as garden-based learning. This approach is strikingly different from those initiatives pursued historically in developing contexts in Africa, Asia, and South America, which, with little success, have aimed to address school food security issues for the rural poor. In this section, we explore these differences.

The main distinction between approaches to school cultivation can be illuminated through the terminology used to name them. "School farming" has prevailed in developing contexts, while "school gardening" has been more common in Europe and North America. Popular understandings of the difference between the two seem to turn on aspects of scale and intent. Whereas "gardening" is generally done on a smaller scale as a pleasurable activity and for personal consumption, "farming" is generally understood to take place on a larger scale for the production of goods to sell or trade. In this paper, we use the term "school cultivation" to refer to both types of activity. Our aim in doing so is, by resituating planning and policymaking for both gardening and farming into the same discursive realm, to confront the "double standards" (Farmer & Gastineau, 2009, p. 155) set for school cultivation among the rich and the poor and to acknowledge the global order that has contributed to this curricular divergence.

In more privileged national settings like the United States, school cultivation has taken the form of "garden-based learning," an approach that emphasizes the development of the whole individual through a cultivation curriculum. In the United States, garden-based learning has some of its strongest roots in the United States Department of Agriculture's youth organization 4-H, which since the beginning of the twentieth century has aimed to develop the "heart, head, hands, and health" of millions of young people in America, who "learn to do by doing" (National 4-H Council, 2010). Since the 1990s, the practice of garden-based learning has expanded to address increasing disparities in wealth and access to healthy food and to address "lifestyle issues" among the urban poor and the suburban middle class. Specifically, these efforts have aimed to counter the alienation of urban and suburban children from nature and agriculture, to promote environmental and agricultural sustainability, to encourage children to make better nutritional "choices," to "teach the joy and dignity of work," and to "green" school grounds through experiential education (Desmond, Grieshop, & Subramaniam, 2004). In California alone, where the state's Department of Education launched a "Garden in Every School" program in 1995, 3000 school garden programs are underway.

Some common assumptions embedded in garden-based learning are illustrated in the quote below, derived from a description of school garden activities in the United States:

> In school garden programs that grow edible produce, students generally learn science and nutrition concepts relevant to growing food while they work in the garden. Students harvest the vegetables and, in some programs, learn to cook nutritious meals from the harvest. Some programs include a "farm-to-school" component in which the school purchases produce from local farmers for its lunch program, and students visit farms to understand where food comes from and how it is grown...In food-growing garden programs, one central health-related goal is to stimulate youth—so many of whom subsist on diets heavily based on packaged foods—to increase their consumption of fresh produce. Students also get some exercise as they engage in weeding, digging, and other manual labor associated with garden maintenance. (Ozer, 2007, p. 847)

The school garden mandate in this context is to expose children to healthier "farm-fresh" foods, to which they usually do not have access, to expose them to cooking (an activity assumed to be foreign to them; see Lalonde, this volume), and to "get some exercise" to counteract what is assumed to be their usual sedentary lifestyle.

In sub-Saharan African contexts like rural Tanzania, however, concerns about packaged foods, a lack of opportunity for food preparation, and a lack of opportunity for exercise are not relevant rationales for school cultivation. Rather, as we illustrate below, school cultivation activities have centered primarily on what is known as "school-based food production," with a secondary emphasis on agricultural education and on the cultivation of national values, with little or no emphasis on other learning objectives. Despite this unity in general purpose, these aims have been put to a variety of political and economic ends in colonial, post-independence, and (in the Tanzanian case) post-socialist eras of schooling. Below we return to the Tanzanian context as an example of the long and winding history of school cultivation on the African continent.

Today, in an era marked by a global policy mandate for Education for All (http://www.unesco.org/en/efa/efa-goals/)—when access has increased, but enrollment and class size have skyrocketed—the cultivation of school food to meet the breakfast and lunch needs of students meets with very little success. School gardens can produce only a fraction of the quantity of food needed by students for meals at school. Misappropriation of the harvests of school gardens also has been an issue. The Food and Agriculture Organization states that misuse of school harvests and student labor has been relatively common, noting that "in the reality of most rural schools, economic concerns often take precedence over teaching objectives, as poorly paid and unmotivated teachers are tempted to use the proceeds...for themselves" (FAO, 2004a, p. 10).

So if the old model of school farms contributes little to its intended aims, what educational, political, and economic purposes have school farms served in Tanzania? What can and does school cultivation accomplish in terms of its educational, political, and economic aims? To answer this question, we contrast the international policy vision of school gardens with the history and current practice of school cultivation in Tanzania.

The New School Garden

In parts of the world facing chronic hunger and malnutrition issues, school farms are being re-conceived as school gardens and partnered with school lunch programs to form a new key component of rural food security policy. In Tanzania, the World Food Programme requires the cultivation of a school farm as part of the community contribution to its Food for Education school breakfast and lunch program. The Tanzanian Ministry of Education and Vocational Training, meanwhile, has expanded the required size of school farms, and village governments have increasingly enforced this mandate, often relocating families who find their homes and farms inside the boundaries of the newly revised borders (a common practice noted by Phillips in Singida region between 2004 and 2006). The U.S.-based youth organization 4-H has launched and is expanding a school garden program in Tanzania. And with the increase in private education models imported from Europe and North America, garden-based learning techniques are the latest trend in elite urban

Tanzanian contexts (the Hekima Waldorf School in Dar es Salaam is one example).

An international consensus on the effectiveness of school cultivation for student learning is reflected in the United Nations' Food and Agriculture Association's endorsement and their recently published Concept Note (2004a) on school gardens. The FAO vision emphasizes school gardens' potential to serve as learning laboratories and as part of a long-term strategy to address food insecurity through improving basic education around issues of nutrition and the environment and introducing improved agricultural techniques of sustainable food production. This vision builds on the garden-based learning model to center learning objectives, while also maintaining that gardens can contribute in long-term ways to school and community food security. The FAO school garden vision is centered on a value of active learning, represented in a quote from Confucius that heads the FAO school garden website: "I hear and I forget, I see and I remember, I do and I understand" (FAO, 2010).

The new school garden, according to the FAO, should be a cultivated area near a primary or secondary school that can both assist learning and produce food and/or income for the school. It should:

> help students learn about food production, nutrition and environment education and personal and social development related with basic academic skills (reading, writing, arithmetic) while generating some food production to supplement school feeding programmes. (FAO, 2004a, p. 4)

Specifically, the FAO argues that the garden should address the following learning objectives: increase the relevance and quality of education by introducing important life skills; help students to start and maintain home gardens; encourage the consumption of fruits and vegetables; offer active learning experiences by linking gardens with math, science, reading and writing; attract children and families to school by addressing topics relevant to their lives; improve attitudes towards agriculture and rural life; teach about environmental issues and nutrition; and give students survival tools for times of food shortage (p. 5). In addition, a school garden should also familiarize students with methods of sustainable agriculture, promote income-generation activities, improve food availability and diversity, enhance the nutritional value of school meals, reduce the incidence of malnutrition among schoolchildren, increase attendance, and compensate for the loss in the transfer of life skills from parents to children due to the impact of HIV/AIDS and increase in child-headed households (p. 5).

The FAO's (2004a) Concept Note qualifies these ambitious aims by noting that the learning objectives of school gardens must be supported by adjustments in the national curriculum, the production of training and curricular materials, teacher professional development, and sufficient funds for these endeavors. It also acknowledges that the economic contribution of school gardens may be minimal. The significant economic and nutritional effects of school gardens, according to the Concept Note, will be in their "multiplier effect" as students learn academic subjects and life skills and help their parents and families start home gardens that diversify and enhance agricultural production (p. 8). Though the Concept Note is ambitious in its educational objectives, its overall aim appears to be to fill the knowledge and skill deficit of rural people without questioning the political, social, and economic orders that construct and produce those deficits in the first place.

The Historical and Current Context of School Cultivation in Tanzania

The idea of producing food in schools and learning about food production in schools symbolically resonates with the concern for food security and sustainable development that characterizes donor concern in sub-Saharan Africa. Yet neither the literature on school cultivation nor our research-based understanding of what is happening in rural Tanzanian school farms indicates that such an educational vision will be easily adopted in rural schools, and if it were, that it could alter the conditions of hunger. In Tanzania the school farm (*shamba la shule*) has long played an important role in most primary and secondary schools. These school farms were not integrated into the curriculum. Rather than supporting student learning, school farm activities in Tanzania have tended to serve the economic ends of producing harvests that would supplement school meals. During the postcolonial period of Tanzanian Socialism (1967 to circa 1985), they also served a political ideology concerned with bridging the divide between an educated non-farming elite and an uneducated agricultural populace. In the following section, we review this history of the school farm.

School Cultivation in British Colonial Schooling (1920–1961)

In colonial Tanzania (then called Tanganyika), school farms emerged in conjunction with British schooling.[2] Schools relied on farms tended by their students for some, if not all, of the food served to boarding and non-boarding students. The colonial government mandated farming activities in government schools as a means of school self-support, whether in the form of food produced for student meals or funds raised by the sale of produce, livestock, or other cash crops that could support the running of the school. The government subsidized such produce with foods like palm oil and milk. Colonial school farms in Singida included the cultivation of crops such as sorghum, millet, cabbage, tomato, eggplant, cauliflower, and papaya ("Memo," 1956). In addition to producing food for the school, students also learned agricultural techniques by doing (*kwa vitendo*) that administrators hoped students would take home with them. The school farm and the teacher farm were both to serve as a *shamba-darasa*—a farm classroom used to model modern farming. A colonial communication noted:

> Because the school farm is for the purpose of lessons, it should be located close to the school. And the children should be taught especially Agriculture from first until fourth grade. The farm of the teacher should also follow the principles of good agriculture so that the children see that their farms and the teacher's farm have the same agriculture. ("Memo," 1956)

Logs were expected to be kept for the farm that would track each agricultural method and the success associated with it. Such data included the dates of planting and second and third weeding, the date of harvesting, any detrimental factors that affected output (birds, animals, drought), and the measurement of the harvest obtained by each class and level ("Memo," 1956). Cattle husbandry was also common. In successful years, the school farm also yielded a profit for the colonial government, for a portion of farm profits was paid out to the colonial government in taxes.

Colonial education policy required participation on school farms and in gardens not only from students, but also from the surrounding school community, which was generally expected to contribute labor and manure ("Community Help," 1954). Yet there were strict rules that no labor was to be misappropriated for the personal gain of teachers. A memo from the Supervisor of Schools in the District Office of Singida to all head teachers addressed this issue, suggesting that such misuse of student labor was already occurring: "Making the students farm for the benefit of the teacher and not for the school, or renting the students out to do work for other people is not right, not even a little, and it is strictly forbidden" ("Memo," 1956). School cultivation, such reprimands communicated, was to benefit the British Empire, not shrewd teacher-entrepreneurs who were now successfully grasping the principles of production for profit that the British had worked so hard to inculcate (Phillips, 2009a).

In the colonial system, the few students who were able to complete schooling generally found themselves in positions of power and privilege with relation to their unschooled counterparts: as colonial administrators, schoolteachers, or church leaders. School farms, as post-independence leaders would point out, were cultivated during the colonial era as part of an education that served the colonial government and "the educated few" ("Summary Report," 1967), not the community or the nation's many.

School Cultivation and Tanzanian Socialism (1961–1982)

In 1961, Tanzania achieved its independence peacefully under the leadership of Julius K. Nyerere, a former teacher who, at independence, became Tanzania's first president. By 1967, Nyerere was leading Tanzania along a new political and economic path of Tanzanian Socialism, or *Ujamaa*. Nyerere (1967) regretted the legacy of colonial education in his new independent nation:

> The education now provided is designed for the few who are intellectually stronger than their fellows; it induces among those who succeed a feeling of superiority, and leaves the majority of the others hankering after something they will never obtain. It...can thus not produce either the egalitarian society we should build, nor the attitudes of mind which are conducive to an egalitarian society. On the contrary, it induces the growth of a class structure in our country. (p. 276)

Given these concerns, Nyerere sketched out a new vision for education in independent Tanzania with his Education for Self-Reliance Policy (*Elimu ya Kujitegemea*). Education for Self-Reliance was embedded in a new political climate in which the percentage of the GDP dedicated to education more than doubled, going from 2.7% to 5.7% (Buchert, 1994). The policy rested on a notion of school students as some of the nation's healthiest and strongest citizens, who not only failed to contribute to national development through their labor, but were also consumers of the labor of older and less privileged people (Nyerere, 1967, p. 4).

In addition to concerns about school students' lack of contribution to national development during their education, Nyerere also worried about the social and experiential separation—produced by the colonial educational system—between an educated elite and those who "fed" them (through agriculture). The British system had privileged knowledge that was largely irrelevant to most Tanzanians and devalued that which a student could learn from his elders. By relying on strong local farmers as supervisors and teachers and making use of agricultural officers, Nyerere argued, "we shall be helping to break down the notion that only book learning is worthy of respect. This is an important element in our socialist development" (Nyerere, 1967, p. 5).

The school farm, for Nyerere, offered an opportunity for the nation to recover some of the national investment in these young people. He argued that teachers and students must be members of a social unit that parallels the family structure.

> And the [school] must realize, just as [the family does], that their life and well-being depend upon the production of wealth—by farming and other activities. This means that all schools...must contribute to their own upkeep... Each school should have, as an integral part of it, a farm or workshop which provides the food eaten by the community, and makes some contribution to the total national income. (p. 5)

Nyerere also emphasized the experiential learning opportunities for agricultural innovation embedded in the school farm:

> [O]n a school farm pupils can learn by doing. The important place of the hoe and of other simple tools can be demonstrated; the advantages of improved seeds, of simple ox-ploughs...The properties of fertilizers can be explained in the science classes, and their use and limitations experienced by the pupils as they see them in use. The possibilities of proper grazing practices, and of terracing and soil conservation methods can all be taught theoretically, at the same time as they are put into practice; the students will then understand what they are doing and why, and will be able to analyse any failures and consider possibilities for greater improvement. (p. 5)

One teacher interviewed in Arusha Town remembered Education for Self-Reliance as it was practiced in her peri-urban village, and the technical support that was available in the post-independence period:

> When I was in school my school had a farm. We sold beans to buy pencils and supplies for school. Vegetables and fruits were grown. Two times each week I worked out on the farm as part of the 45-minute agricultural lesson. Schools in town where I studied got water from taps, but in villages students filled water in a river. The schools boiled corn for school lunch. Parents bought vegetables and fruits for not expensive prices. I made a natural pesticide which was soap and pepper put together overnight and put on the leaves of plants. We ate corn and beans each day for school lunch, and vegetables and meat once a week. It was nice to have an agricultural officer in villages who specialized in schools.

Through Education for Self-Reliance, primary school enrollment rose to 93% by 1980; adult literacy from 10% to 90%; life expectancy from 35 to 50 years, and by 1980, 79% of high-level civil service positions had been "localized," that is, transferred from British officials to Tanzanians. Yet not all post-independence officials shared Nyerere's enthusiasm for active learning as the primary motivator for school cultivation. Many saw school farms primarily as sites for the application of innovative agricultural techniques developed elsewhere. Tanzania's Director of Agriculture, for example, emphasized in 1967 that:

> experimentation should be left to people who are qualified to do so. [Education for Self-Reliance involves] teaching the pupils how to obtain the required inputs at the right place, right time, and right quantities. If this were done, I am convinced production would go up and the income would exceed that from salaried employment, thereby creating an incentive for the school-leavers to go back to the land. ("Farming Could Be Lucrative," p. 5)

The visiting Minister of Education emphasized to his Singida audience in 1975: "Education for Self-Reliance is not a 'hobby' nor an experiment, rather [its goal] is production" ("Report of Tour," 1975). Likewise, the success of Education for Self-Reliance was touted in 1968 by Minister of Education Muhaji as "the psychological change of the students' attitude towards manual work" ("New Form," 1968). And the construction of a cattle dip was lauded in a national newspaper for "making students aware of their responsibilities to the community and the nation" ("Students Build Cattle Dip," 1968). During Education for Self-Reliance, school farming and animal husbandry came to be valued by colonial administrators for their economic production and cultivation of "a value of work"—aims eerily similar to those of the colonial administrations (Phillips, 2009b)—not for the learning of academic subjects or the education of a politically active populace.

In many contexts, Education for Self-Reliance came to legitimize the practice of using students as work teams. Teams of students were often "rented out" by schoolteachers to private citizens for work on farms or house construction. The profits, ostensibly, went to the school, but many people we interviewed noted that it was likely such profits were often enjoyed privately by the teacher. Rumors of misappropriation and corruption by village leaders and schoolteachers characterize the memory of Tanzanian Socialism for many people we spoke with in Singida, Lindi, and Monduli districts. That said, the principles of Education for Self-Reliance and the intention of Nyerere are still held up as political and economic ideals by many rural Tanzanians.

School Gardens in the Context of Market-based Reform and Education for All (1982–Present)

By the early 1980s, all of Africa was caught up in the effects of a global economic recession, and Tanzania's macro-economic challenges had intensified. The World Bank and the International Monetary Fund stepped in as lending institutions for Tanzania and other severely indebted countries and declared Tanzania's level of government spending on education and other social services impossible to sustain. In 1982 the Tanzanian government instituted the Structural Adjustment Programme that the international lenders had set as a condition for aid. The values of access and equity were de-emphasized in educational policy (Buchert, 1994). While some economic recovery was demonstrated as a result of structural adjustment, the impacts on education and the rest of the social sector were severe. School enrollment plummeted with the implementation of school fees. Poverty and illiteracy increased. There is little evidence of explicit national policy on school cultivation during this time. The responsibility to feed students, like that of paying fees, fell on families, many of whom withdrew their students from schools due to an inability to pay.

In the wake of the 1990 United Nations "World Conference on 'Education for All'" in Jomtien, Thailand, the goal of universal primary and secondary education became part of Tanzania's broader vision of poverty elimination by 2025. In 2001, the government of Tanzania eliminated school fees for primary schools, and in 2005 the government ordered the construction of a secondary school for every ward (a grouping of 4–6 villages each containing its own primary school) to increase opportunities for education. Both goals were largely realized by 2010, though not without great cost to communities who were largely responsible for fulfilling the policy through mandated contributions of labor and materials. Though enrollment statistics show over 100% enrollment, attendance rates in Tanzanian primary schools suggest that students face other barriers to education. Acute and chronic hunger issues and rote pedagogical style have been noted as significant (Twaweza, 2010; WFP, 2010).

Though little research has been carried out on the contemporary practice of school gardening in rural Tanzanian schools, O-saki and Agu's 1999 study of classroom interaction in four districts has relevance to the major pedagogical issues around school farms. They found that, in general, teaching styles in these schools were orthodox. Teachers tended toward "chalk-and-talk"—lecturing, rather than providing the opportunities for active learning that proponents of school garden curricula tout. School cultivation activity cited focused mainly on watering gardens and harvesting produce—with no attempt to integrate the activities into the academic curriculum (O-saki & Agu, 2002, pp. 108–109).

Such tasks, O-saki and Agu quietly assert, "smack of child labour" (2002, p. 109). O-Saki and Agu noted that in the rural schools, issues were more severe: "teachers took class work less seriously and focused more on using the children as cheap domestic and school labour, which depleted the children's energy...And where girls did such work for most of the school time, it eliminated their motivation" (p. 113). Some children even complained about the teachers' "failure to account for income and expenditure (relating to school self-reliance harvests produced by the children)." It was not simply that their school farm work contributed little to their learning, but that in many cases their labor did not even contribute to their diet. O-saki and Agu concluded that "Some teachers may be wrongly interpreting [Nyerere's] Education for Self-Reliance philosophy in relation to education and work" (p. 115).

In summer 2010, we followed up on these themes raised by O-saki and Agu, conducting interviews and focus groups in two schools in Rural Lindi, one school in rural Monduli district, and one school in Arusha town. We spoke with teachers, villagers, and community members about school cultivation in terms of what they referred to as "school farms" (*mashamba ya shule*)—larger plots of land planted with staple crops like maize and beans—and "school gardens" (*bustani za shule*)—smaller plots planted mainly with fruits and vegetables that often require watering by hand to supplement seasonal rains.

We found that only the Arusha town school had an active school garden or farm, sponsored by 4-H, who offered technical assistance and curricular support. One of the schools in Rural Lindi had planted orange and palm trees and sold their fruits and, rather irregularly, planted and harvested maize and beans. All of the rural schools cited as reasons for their lack of garden or farm activities challenges with recent droughts or irregular rains, lack of access to water for irrigation at the school, pests such as birds or livestock destructive to crops, teachers' lack of capacity to take on a school cultivation curriculum, and/or inadequate time in the school day for school cultivation activities given the demands of preparing for national examinations. All sites' school meal programs benefitted from school cultivation, but they also acknowledged that harvests did not contribute significantly to a meal program. One kindergarten teacher in Arusha noted:

> It is quite unadvisable for kids to grow food for lunch as in the past because these days schools have no farm land. In the past kids grew food for lunch though the food lasted for quite a short period. As I have said before, these days schools have not enough land to farm and this makes a big change. The best thing is for the government to contribute to schools and parents as well if they really need kids to have lunch rather than giving cabinet members such big salaries, luxurious cars and so many other incentives for nothing.

All in all, village officials, teachers, and community members in all three regions generally associated school cultivation with the production of school food or cash crops that would financially support the school. As one Lindi parent noted:

> We don't have this plan of farming for learning. We farm for the purpose of school funds only. This plan of encouraging learning has been started this year through the

> government policy of "Agriculture First," but we haven't started it yet. We think we will start next year because we are already late.

In the rural Arusha school, farms were sometimes being used to teach *kwa vitendo*—agricultural or biology education using active learning techniques. One Arusha teacher noted: "for example, when we teach the fertilization of flowers we send the students out to look at the flowers...they see the ravines and the difference when there is a drought and when they cut trees."

In general, however, what people could "learn by doing" through school cultivation was generally understood by teachers and parents to be only agriculture; for example, students could learn about the use of manure and issues such as soil erosion and environmental degradation. There was little recognition that farms and gardens could be used to teach other parts of the national curriculum, like science, mathematics, or writing skills, through active learning techniques. Rather, the notion that a school farm or garden could also be a classroom was limited to people's familiarity with the *shamba-darasa* (literally, the farm-classroom), or the model farm that is used to teach villagers improved farming techniques.

Both teachers and parents perceived that the use of school gardens to teach the curriculum and/or cultivate school food would compete with teachers' time spent preparing students for exams. There seemed to be considerable concern that such activities would interfere with students' performance on national exams, rather than supporting and enhancing students' understanding of material. However, school cultivation that would provide for a school lunch for students, if not too time-consuming, was seen as a worthwhile use of students' time. Parents, teachers, and village officials, though, confirmed that both student labor and the land available for school cultivation are not enough to sustain a school food program.

One recent shift we noted is the renewed government emphasis on agriculture as a means to development. Many cited President Jakaya Kikwete's (president since 2005) new slogan, "Agriculture First," as testament of a return to a more essentially Tanzanian way of life, against the grain of advice from powerful international governments. One Arusha town teacher observed:

> Now agriculture has changed, but the government is encouraging the farmers to start farming like they used to because they see we have lost out. Farmers have become increasingly poor and they have no money and no business, but there are farmers who are agreeing now to return to the ways of the past. Others say forget it, we've had enough. The government is persuading them by giving help that might help them return to our initial agriculture. Because CCM [the ruling party] is saying "Agriculture First." This is similar to Nyerere's time, when he insisted "Agriculture is the backbone

> of the nation." This is true because each person depends on agriculture: every person needs to eat, for without eating there is no life.

Some teachers and parents we spoke with expressed concern about the perceived return to agricultural education that a renewed emphasis on school cultivation might represent. In Arusha, elders and community members had far less interest in their children learning agriculture than their learning the national curriculum. Their main interest was that their children perform better on national examinations so that they could have access to opportunities for secondary education. Teachers and parents saw this path as more fruitful than one based solely on farming and herding, for they could "sit in an office and receive a salary."

In the Lindi schools, other "self-reliance" activities were also carried out. Students collected firewood and palm fronds (used to re-thatch house roofs) to sell for the schools' profit. Based on concerns raised in the research and policy literature (FAO, 2004b; O-saki & Agu, 2002), we asked parents if they had any concerns with their children doing "farming work" at school. Most parents had no problem if the time spent was not excessive and if the children themselves would materially benefit from it through school meals or improved supplies and services. One noted,

> The children should get a certain percentage of the money; they can use it to pay for school supplies, shoes, and save for secondary school. It is fair to give kids money (from selling school produce) when they are hungry; otherwise they will learn to steal. The kids should get money, produce, or school lunch. Parents should pay a cheap price to buy produce so parents can buy the produce to feed their families and the school can also get money. School lunch is very important because kids can't learn on an empty stomach.

But many did express reservations about who might benefit from the production of food on school farms, due to possible misappropriation of harvests and profits by teachers. "It is necessary that there is good supervision;" several of them insisted in a focus group interview, "there must be a good plan."

Discussion

Notions of self-reliance have framed diverse political, economic, and social projects in Tanzania's history. These ideas have had significant effects on the forms education has taken and the ends that it has served. During colonial rule, all school learning was oriented toward maximizing economic productivity for the colony, that is, educating citizens to carry out the colonial agenda through a transfer of knowledge, skills, and—ultimately—values that would allow Tanzanians to be productive colonial subjects requiring as few further inputs as possible. Colonial administrators sought for schools to be self-sustaining, and the school farm's role was to contribute to this aim through the agricultural curriculum and through its economic contribution to school self-support. This would be a far cry from self-reliance in the Nyererean (1967) sense: learning to labor was meant to contribute to an overall surplus for the Empire's profit and expansion.

During Tanzanian Socialism school learning was also geared toward economic productivity, but toward a very different notion of self-reliance. Self-reliance for Nyerere (1967) meant national sovereignty (independence from world economic powers) and the achievement of modern services (health care, education, electricity, etc.) through small-scale action and organization. Such self-reliance would grant Tanzania true—not simply nominal—political and economic independence from former colonial and other world powers. A challenge to this objective became keeping up with, or restructuring, the aspirations, consumer tastes, and value of economic, political, and educational hierarchy developed during the colonial period. Nyerere's Education for Self-Reliance tackled such challenges directly through an educational emphasis on the development of civic virtue, egalitarian sentiments, and a value of national social cohesion and identity. Collective work on school farms, specifically by those en route to privileged government employment (i.e., students), provided the mechanism through which these shared values could be developed. Yet the actualization of this project often took very colonial forms—emphasizing the value of work and learning as "banking education" (Freire, 1970)—the transfer of knowledge and skills from an educated and political elite to the common man. In practice then, labor was emphasized over learning.

In this first decade of the new millennium, learning is still aimed toward national economic productivity but also toward the development of a democratic market economy. There is a tension here between understanding education as a "human right" (as it is framed in Education for All) yet one that is still accessed and differentiated through market forces—through, for example, the capacity to purchase school uniforms and pay mandatory school "contributions"; to access the social and political capital necessary to gain a secondary school placement where teachers, books, and training are sufficient; or to pay private school tuition for an education that offers the linguistic training needed for promotion. "Self-reliance," a concept still invoked today by politicians, now stands more for a neoliberal emphasis on individual responsibility and pulling oneself up by one's bootstraps than as a postcolonial cry of self-determination and independence. Self-reliance and the cultivation of school farms and gardens have been incorporated into an educational development discourse on sustainability. Yet the aim of such sustainable development initiatives as school gardens seems to be that rural African communities become productive self-supporting entities requiring as few further inputs from outside forces as possible—an objective not all that different from that of colonial education. One elder we spoke with in Lindi region articulated his frustration with self-reliance as a political objective in today's world: "Self-reliance has its good points, but its bad points too. Because if you say you are going to depend on yourself, but there's no certainty that you can, then it is no good." Rural Tanzanians are painfully aware of being subject to and at the whim of market forces at the same time as they are continually asked to stand on their own two feet and "help themselves."

A central concern that has emerged in the literature on school cultivation in developing contexts is that of child labor, a subject we would like to briefly address here. Anthropologist Kristen Cheney (2007) offers an important critique of the ambiguity of notions of childhood in postcolonial Africa. She writes that the global rights discourse constructs an understanding of the "universal child" that is typically informed by Western values and seeks to "free children from the negative constraints of their own traditional cultures, often seen as negative and 'antimodern'" (p. 44). Education for All, likewise, sets children apart from their own communities where they "might otherwise be integrated in vital ways—in the labor force, for example" (p. 44). Yet, she goes on to note, "postcolonial economic decline only entrenches persistent African cultural notions that children are essential family resources, not individuals endowed with rights and freedoms independent of family and community" (p. 56). The parents we spoke to are indeed pragmatists; their children must learn sooner rather than later to support themselves and contribute to the livelihood of their family. This seems an important parental value to respect in the conceptualization and implementation of school cultivation initiatives. Our hope is that strengthening the political position of students and their parents with respect to teachers, and of rural schools with respect to a government sworn to serve them, will curb misappropriation of children's labor and school resources. Students should gain from their educational labor and a laboring education in both material and intellectual ways.

Achieving these gains, we argue, can occur through development interventions such as the development of school cultivation programs, but not necessarily in the form in which they are currently conceived. The FAO's school gardens, for example, center the dual objectives of, first, agricultural improvement through the transfer and application of technical knowledge and, second, nutritional enhancement through learning the principles of nutrition and cultivating new tastes. Education is also emphasized but is connected more to the goal of certification and the pragmatic realities of passing national examinations in an age of competitive education than creating educated persons and citizens. Local and life-relevant skills are also emphasized, but a familiar overall objective remains: filling the knowledge and skill deficit of rural people without questioning the political, social, and economic orders that construct and produce those deficits in the first place.

FAO reports themselves acknowledge the inadequacy of technical development, and the importance of "empowerment":

> The "agriculture-only model of rural development" has proven inadequate to address poverty reduction, rural development and sustainable natural resources management. The latest thinking and good practices in such domains indicate that the *empowerment* of poor people, policy and institutional reforms in the rural sector leading to participation of stakeholders needs to be the starting point...(Gasperini, 2000, para. 5; emphasis added)

Yet despite this acknowledgment, discussions of *power* in development processes are still avoided. The FAO exemplifies this contradiction in policy circulars such as the School Gardens Concept Note. There is no indication of how "empowerment" can and will take place through its vision for garden-based learning. It is understandable that international organizations do not want to address political issues that can be seen to impede on national sovereignty. Yet as nongovernmental and civil society organizations pointed out during the 2002 "Forum for Food Sovereignty," not only is there a lack of political will to combat hunger, but also

> too much political will is used to promote policies that actually exacerbate hunger...It is clear that strategies to overcome or reduce hunger, malnutrition and rural poverty need to both promote new policies as well as challenge the national and international policy environment that hinders access to productive resources or to an income sufficient to feed oneself. (Windfuhr & Jonsen, 2005)

We argue for the need to situate school cultivation initiatives within processes that, rather than simply incorporating local decision-making processes into an end or trajectory pre-established by the global development industry, instead build on a dialogical process between teachers, agricultural experts, students, community members and leaders. This process should situate planning for school cultivation and rural development in (1) a historical contextualization of school gardens in African contexts and (2) a critical appraisal of local and global systems of food production, distribution, and consumption. Questions to be debated within communities could include:

- As rural food producers, how and why are we often the first to go hungry? (Phillips, 2009b; Shipton, 1990). What are the economic, historical, and political conditions that construct us as "food insecure" in the first place?

- How can we use school gardens to not only disseminate and practice technical knowledge, but to reshape the conditions of our poverty?

Such a critical approach to addressing issues of poverty and democracy as they relate to world distribution of food should not be limited to school cultivation initiatives in developing contexts. Rather, in this global era in which it is clear that we occupy not different food worlds, but the same one, school cultivation initiatives in the Global North should be embedded in such a food sovereignty framework as well. For, in Paulo Freire's words, education can either function "as an instrument that is used to facilitate the integration of the younger generation into the logic of the present system and bring about conformity to it, or it becomes 'the practice of freedom' the means by which men and women" and, we would add, boys and girls "deal critically and creatively with reality and discover how to participate in the transformation of their world" (Freire, 1970, p. 15).

Notes

1. This summer 2010 research consisted of site visits to school communities in Lindi Rural and Monduli districts, using formal (semi-structured), informal, and focus group interviews with parents, village officials, and village elders regarding school gardens.
2. After the German defeat in World War I, a newly formed League of Nations handed control of Tanganyika over to Great Britain in 1920. It was missionaries and the British colonial regime who introduced formal schooling into Tanzania. Britain ruled Tanzania until independence in 1961.

References

Agriculture and Natural Resources Team. (2004). *Agriculture, hunger and food security*. London: Department for International Development.

Buchert, L. (1994). *Education in the development of Tanzania 1919–1990*. Athens: Ohio University Press.

Cheney, K. (2007). *Pillars of the nation: Child citizens and Ugandan national development*. Chicago, IL: University of Chicago Press.

"Community help on school buildings of the Voluntary Agencies: Memo from Augustana Lutheran Mission to Singida District Commissioner." (1954, May 29). Tanzania National Archive E1 68/19/6 III.

Desmond, D., Grieshop, J., & Subramaniam, A. (2004). *Revisiting garden-based learning in basic education.* Rome: Food and Agriculture Organization.

Farmer, P., & Gastineau, N. (2009). Rethinking health and human rights: Time for a paradigm shift. In M. Goodale (Ed.), *Human rights: An anthropological reader* (pp. 148–166). West Sussex: Wiley-Blackwell.

"Farming could be lucrative." (1967, April 13). *Standard.* T A3138 #92, p. 5.

Food and Agriculture Organization. (2004a). *School gardens concept note: Improving child nutrition and education through the promotion of school garden programmes.* Rome: Food and Agriculture Organization.

Food and Agriculture Organization. (2004b). *The state of food insecurity in the world, 2004.* Rome: Food and Agriculture Organization.

Food and Agriculture Organization. (2010). School gardens website. Retrieved October 1, 2010, from http://www.fao.org/schoolgarden/.

Freire, P. (1970). *Pedagogy of the oppressed.* New York: Herder & Herder.

Gasperini, L. (2000). *From agricultural education to education for rural development and food security: All for education and food for all.* Rome: Food and Agriculture Organization. http://www.fao.org/sd/exdirect/exre0028.htm.

Hodgson, D. (2001). *Once intrepid warriors: Gender, ethnicity, and the cultural politics of Maasai development.* Bloomington: Indiana University Press.

Maxwell, S. (1996). Food security: A postmodern perspective. *Food Policy, 21*(3), 155–170.

"Memo from the District Supervisor of Schools to all Head Teachers of Government and Native Authority Primary Schools, Singida and Iramba." (1956, July 31). Tanzania National Archive E1/10 #33, p. 2.

National 4-H Council. (2010). "4-H Home Page." Retrieved October 1, 2010, from http://www.4-h.org/.

"New form: Public will be given voice." (1968, August 13). *Standard.* TNA ACC 584 EDG #73, p. 5.

Nyerere, J. (1967). *Education for self-reliance.* Dar es Salaam: Government of United Republic of Tanzania.

O-saki, K. M. & Agu, A. O. (2002). A Study of Classroom Interaction in Primary Schools in the United Republic of Tanzania. *Prospects XXXII*(1), 103–116.

Ozer, E. J. (2007). Effects of school gardens on students and schools: Conceptualization and considerations for maximizing healthy development. *Health Education & Behavior 34*(6), 846–863.

Phillips, K. (2009a). *Building the nation from the hinterlands: Poverty, participation and education in rural Tanzania.* Ph.D. Dissertation, University of Wisconsin-Madison.

Phillips, K. (2009b). Hunger, healing and citizenship in central Tanzania. *African Studies Review, 52*(1), 23–45.

"Report of Tour of Minister of Education Comrade I. Elinewinga from Dec 5 1975 to Dec 18 1975." (1975). Tanzania National Archives EDN 20/106.

Sen, A. (1981). *Poverty and famines: An essay on entitlement and deprivation*. Oxford, England: Oxford University Press.

Sen, A. (1999). *Development as freedom*. Oxford, England: Oxford University Press.

Shipton, P. (1990). African famines and food security: Anthropological perspectives. *Annual Review of Anthropology* (19), 353–394.

"Students build cattle dip." (1968, August 2). *The Nationalist*. Tanzania National Archives ACC 584 EDG #57, p. 8.

"Summary report of Secondary Teachers' Conference Tanzania XIII at Machame Girls' Secondary School." (1967, April 27–29). Tanzania National Archives ACC 584 A3/38/I.

Twaweza. (2010). Malnutrition: Can Tanzania afford to ignore 43,000 dead children and Tshs 700 billion in lost income every year? Dar es Salaam: Uwazi InfoShop.

United Nations. (1975). *Report of the World Food Conference, November 5–16, 1974*. Rome: United Nations.

Via Campesina. (1996). "The right to produce and the access to land. Food Sovereignty: A Future without Hunger." Retrieved from http://www.voiceoftheturtle.org/library/ 1996 Declaration of Food Sovereignty.pdf

Windfuhr, M. & Jonsen, J.. (2005). Food sovereignty: Towards democracy in localized food systems. Bourton-on-Dunsmore, Rugby, Warwickshire: ITDG Publishing.

World Food Programme. (2010). Country Programme–United Republic of Tanzania 10437.0 (2007–2010). Rome: World Food Programme.

• CHAPTER FOUR •

Defining the "Problem" with School Food Policy in Argentina

Sarah A. Robert
Irina Kovalskys

Argentina's school and public health professionals are both concerned about school food policy. Yet these groups have decidedly different understandings of the "problem" with school food. Both school and public health professionals agree the current Promotion of Social Nutrition Program (*Programa de Promoción Social Nutricional*, PROSONU) does not meet a minimum goal of providing a healthy nutritional foundation for the nation's future citizens (Buamden, Graciano, Manzano, & Zummer, 2010; Robert, 2010). Each group's definition of the "problem" the current feeding policy fails to address, however, is quite different based on the nutritional issues that each group foregrounds, as well as the approaches they take toward fixing the issue. Teachers and school administrators who have daily contact with students and frequent contact with families struggling to make ends meet consider hunger–a deficiency in caloric intake–and food insecurity–a lack of consistent resources to purchase, barter, trade, or produce food–the problem. Their understanding of hunger is based on a contextualized assessment of the everyday lives of their students. On the other hand, public health professionals, including medical doctors, gain access to schools to conduct nutritional health assessments because clinical data suggest a growing percentage of youth are obese and are developing forms of diabetes at earlier ages. From their perspective, the primary public health problem is malnutrition–sufficient caloric intake, but from food lacking in nutritional substance needed by youth to grow healthy. These varied understandings of the nutritional needs of youth lead both school and public health professionals to act on Argentina's school food policy ecology differently and to act independently rather than cooperatively.

For this chapter, we placed three studies of school feeding—one from educators' perspectives and two from public health perspectives—into conversation toward developing a nuanced ecological analysis (Weaver-Hightower, 2008) of Argentina's school food policy. We first situate both groups within the broader context of Argentina at the beginning of the twenty-first century to clean up some of the messiness inherent in a policy ecology. We then discuss each group's impetus for acting on the policy process, definition of the policy problem, and proposed changes to school feeding. Then, we bring the studies into conversation to attempt to reconcile diverging definitions of the policy problem and forge a stronger, cooperative vision of what reform is needed. Our findings suggest that both groups' pressure to amend school feeding policies does indeed address shortcomings, but different ones that each only consider half of the school-feeding problem. Nevertheless, our findings reveal that the different amendments for improving the policy are compatible. A broad spectrum of nutritional needs is present in Argentina's public schools, from hunger to malnutrition, and reconciling the multiple definitions of hunger and, then, forging a broader definition of the problem is needed for school food policy reform.

We argue for conversations as one way to overcome the current fragmentation and entropy of the system, shifting relationships from quasi-symbiotic to collaborative. Collaborative policy relationships formed from holistic, ecological models of policy analysis can potentially lead to feeding programs that reflect the needs of Argentina's youth at the beginning of the twenty-first century: a need to improve the nutritional health, nutritional health education, and therefore educational opportunity of all youth. We warn that concerned groups of actors often talk in isolation from each other or, worse, past each other as they push for change in a policy environment that is fragmented by decentralization of social services. We argue that as part of conversations among educators and public health professionals, both groups must acknowledge the multiple meanings of hunger as employed in everyday life and collaboratively redefine the school food problem accordingly. We call for more participatory policymaking to continue to confront hunger and malnutrition in schools. We suggest that collaborative policy dialogue also may lead to policy change that contemplates an often-forgotten goal of even having a school feeding program in the first place: social justice, or the right of all children to sufficient nutrition and an education.

The case of Argentina is important to examine because it reflects the nutritional concerns of a developed and developing nation. Hunger and malnutrition coincide in Argentina's vast educational system, with a school-feeding program suffering from the challenges of high rates of poverty and the ills of the modern food system in a decentralized context of governance. Furthermore, Argentina boasts extensive educational and healthcare systems reflective of developed nations but with the funding and infrastructure needs of a developing nation. Beyond the national case, the studies at the foundation of this endeavor reflect the epistemological, methodological, and institutional chasm between different groups—educators and public health professionals—confronting childhood and adolescent nutrition in many nations' public schools. As budgets to fund social services such as school feeding continue to be cut, that chasm must be bridged to ensure that students' nutritional and educational needs and rights are fulfilled.

Toward a Cooperative Relationship and Dynamic Policy Ecology

According to an ecological framework for analyzing education policy, multiple actors interact within a complex, multi-tiered environment to shape food policy for schools. We examine Argentina's current school food policy from the perspectives of two groups of actors: school officials and public health officials. School professionals work on the ground in school environments. The boundaries of their policy work are often limited to the needs of their students. They care about youth beyond their school's walls, but the impetus for their agency is their students' well-being. While they may travel beyond the school community to press for change, their understanding of the food problem is drawn from their knowledge of the current policy shortcomings perceived in their students' lived experiences as well as their own. Public health officials also enter schools. They conduct short-term analyses of students' nutritional health. Their work is bounded by the data collected during visits but also influenced by extant conditions that led them into schools in the first place, including a growing concern for obesity and diabetes in Argentina's adolescent population (Kovalskys, Rausch Herscovici, & De Gregorio, in press; Mazza, Ozuna, Krochik, & Araujo, 2005).

Both groups are working within a system that suffers from the *entropy* of an outdated and fragmented program. Education has expanded dramatically since school food policy's emergence. However, the feeding program's objective continues to be to provide a basic nutritional boost to the poorest and youngest students. School professionals have pushed for *adaptation* of the current policy. Teachers and directors act on the existing policy, pressuring for an extension to include feeding all students at their high schools. Their pressure on the system at the beginning of the twenty-first century reflects the decentralized nature of educational services that places responsibility (and blame) on educators for the educational and related social needs of their students.

The current feeding program reflects qualities of policy entropy in relation to the changing nature of food and nutrition that has occurred over the last sixty years (Nestle, 2007). No longer is the nutritional concern of the poor isolated to a lack of food. With the emergence of industrial food products that are cheaper to produce, distribute, and purchase, the food "problem" of the poor is a lack of nutrients and an excess of empty calories; obesity is an epidemic among the poor for the first time in the history of humanity (Peña & Bacallao, 2000; Drewnowski, 2004). Public health officials *anticipate* a rising public health concern of adolescent diabetes and obesity. Research of school food supports that concern, with studies revealing that the food offered students during the school day does not satisfy nutritional needs (Buamden et al., 2010). Public health officials propose that policy mandate healthier food and nutritional education so students eat and choose to eat healthy for life (Subcomisión, 2005).

School and public health professionals are working within the same national policy terrain, yet the nature of their relationship is tenuous. In the study that follows we illustrate that relationship by elaborating each group's involvement in changing Argentina's school feeding program as the work of those groups currently exists in context: separated. However, by weaving the two into this conversational chapter, we illuminate potential for collaborative relationships. School and public health professionals may coexist and work toward a common goal (Weaver-Hightower, 2008, p. 156). Using an ecology metaphor for school food policy analysis calls for probing the complexity of the problem, potentially bringing to the fore insights of the multiple actors acting on behalf of youth in the national context.

To develop a cooperative relationship and dynamic policy ecology, we examine each group's understanding of the nutritional "problem" and each group's approach for addressing it. What is the problem (e.g., definitions reexamined)? How do we know it is a problem (e.g., ways problem is analyzed)? How is the problem addressed (e.g., solutions offered)? The conversation that transpires as the two approaches are compared and contrasted starts a broader conversation regarding the health of adolescents and the role of school feeding in public schools.

Over a Century of School Feeding in Argentina, 1900–2009

School feeding is part of a complex ecology that inherently involves diverse, concerned groups of actors in the "messy workings of widely varying power relations, along with the forces of history, culture, economics, and social change" (Weaver-Hightower, 2008, p. 154). The history of Argentina's school food policy is no different, involving change in who is served, what they are served, and why, as well as what the program is called, which governmental organization manages the program, and where funding comes from, if any is provided. Tracing out the organization, focus, and control of the program over the past century provides a historical backdrop to the education and public health studies that follow, illustrating the ways that both groups act within a rich historical context that has shaped teachers' definitions of what school feeding in the nation can and should do. The same context also implicates public health officials—medical doctors, specifically—in the decision-making process of feeding the nation's children.

Early in the twentieth century, schools were linked to public health concerns for meeting future citizens' health and nutritional needs. Argentina's population, like the United States, was increasing with an influx of immigrants, mostly from Europe. Mass education begun in the 1880s continued to expand to provide for an increasing population. At the same time, the medical "sciences" were quickly evolving. Medical doctors were assigned to schools as part of the School Medical Corps (*Cuerpo Médico Escolar*). An official curriculum with roots in nation-building struggles (Puiggrós, 1990) accompanied the more commonly known school curriculum that imparted basic literacy and math skills; immigrant and Argentine-born students of European descent were to be educated about hygiene and other regulatory practices related to controlling one's body and controlling society (Lionetti, 2007). (It is important to note that excluded from education and full citizenship were indigenous children, children of African descent [*negros/as*], orphans, and other exploited populations [Puiggrós, 1990], which is why those included in the official educational project are qualified.) Though the dominant model of citizenship derived from liberalism and positivism was contested by groups as diverse as Conservative Catholics and Anarchists, the official curriculum demanded a pedagogical subject that was obedient to the nation and accepting of unequal economic relations (Puiggrós, 1990). Creating a healthy population that would grow into a loyal and economically productive Argentine citizenry was an important goal of the school-medical link. Equally important to underline is the fact that schools were battlefields where struggles for control of Argentina and its people were fought.

Public education became entwined with public health toward an ultimate goal of producing healthy, economically productive Argentine citizens. Feeding students was considered a form of *social assistance* that, though distributed through schools, would be done under the watchful eyes of medical doctors (Billorou, 2008). The colloquial "cup of milk" (*copa de leche*) provided public school students throughout the nation with a nutritional supplement as early as 1906 (Billorou, 2008, p. 178; Buamden et al., 2010). Milk and periodically a small bun augmented caloric and protein intake. Weak children (*niños débiles*) (Britos, O'Donnell, Ugalde, & Clacheo, 2003, p. 4) needed calories—particularly from protein sources—for growth. Also targeted were students not making progress in learning and/or not attending school regularly. The number of students served, what they were served—milk and/or bread—and who was responsible for funding the program varied from school to school; the early school feeding program was decentralized, dependent on local politics and preoccupations.

The distribution of a school snack or, in some instances, of more elaborate meals in school cafeterias was scattered and not organized or consistently funded beyond school-level governing bodies until the worldwide economic collapse in the 1930s. In 1936, national funds were established to support school cafeterias throughout the nation with Palacios' Law (*Ley de Palacios, 12.341*). The first municipal cafeterias were opened in the city of Buenos Aires in 1937, followed by the Buenos Aires Province. These provided milk, bread, and in some cases breakfast and/or lunch. School feeding discourse shifted away from a focus on assistance toward a multi-faceted discourse of *social protection*: to protect mothers and children through education and medical advice emanating from notions of scientific motherhood (see Nari, 2004); to protect society from future immorality (Leguizamón, 2005, p. 94); and to promote social collaboration to resolve social problems. While serving the nutritional needs of students, school feeding became closely linked with a need to protect (and educate) poor families toward self-sufficiency.

In the late 1940s and 1950s, school feeding became one of a host of social welfare policies instituted by General Juan Domingo Perón and Eva Perón. As part of a broad social welfare state and the foundation for a politics of *clientelism*, or an exchange of services and goods for political support (normally votes), students continued to receive a wide variety of nutritional support through their schools, an entitlement of future citizens. Management and responsibility for funding school feeding, known as the Directorate for School Help (*Dirección para Ayuda Escolar*), shifts back and forth between directives and ministries, starting off in Public Health and Social Assistance (*Salud Publica y Asistencia Social*), then moving to Work and Provisions (*Trabajo y Previsión*), to the General Directorate of Social Assistance (*Dirección General de Asistencia Social*), and, lastly, ending up in the Ministry of Education and Justice (*Ministerio de Educación y Justicia*). It is during this moment that rights-based discourses first are connected to the school food program, a point we revisit later in the paper.

At the end of the 1960s, the national state established guidelines for managing school cafeterias and for menus to ensure that students' caloric and nutritional needs were met. This symbolically represented the national government's renewed interest in school feeding being connected to nutritional science. Prior to this time, how cafeterias were run, by whom, and what was served was left to the discretion of local governing bodies. The new guidelines instituted the quality and quantity of food influenced by nutritional health research. Students were to receive 1,000–1,200 calories per day from school food. In 1967 the program was piloted in the province of Tucumán. It was extended to 21 of 23 provinces in 1972 under the program name, School Cafeteria Program (*Programa de Comedores Escolares*), and finally expanded to all provinces by 1984 after the return to a democratically elected government. Beginning in 1984, school feeding was once again renamed becoming the Program for the Promotion of Social Nutrition (*Programa de Promoción Social Nutricional*) and managed by the Ministry of Health and Social Action (*Ministerio de Salud y Acción Social*), but run out of schools by school staff.

The early 1990s were characterized by a second wave of neoliberal reform—"the market-oriented restructuring of state social policies" including health and education (Ewig, 2010, p. ix). Under President Carlos Saul Menem's administration, Congress passed the Transfer Law (1992), making provinces and the Autonomous City of Buenos Aires responsible for the management of schools. "The majority of...jurisdictions share similar problems, yet each program acquired different characteristics related to local governments' ability to administer the program" (Britos, 1995, p. 77).

School decentralization accentuated the differing institutional capacities of subnational jurisdictions to administer a wide range of educational programs, including school feeding. Yet decentralization does not seem to change—for better or worse—school feeding. In a study conducted in 2007, school officials responsible for feeding children were found to lack knowledge regarding nutritional needs of students based on age-specific guidelines; Five-year-olds and nine-year-olds receive the same amount of calories (Buamden et al., 2010). Though adequate caloric intake for the youngest, the amount is not a sufficient supplement for older students. Food quality continued to suffer as a result of limited funding. The concerns of the 1960s, 1970s, and 1980s were not addressed.

Decentralization of school administration, but centralization of funding, did not help. The inconsistent transfer of funds for school food programs also led to inconsistent feeding, with the national, state, and even provincial governments accusing local school districts and municipalities of misappropriation of funding. These accusations were picked up in the media and even on the Ministry of Education's Web site. In posted comments on the Ministry Web site between September and December of 2008, authors disputed whether students were being fed and whether the cafeterias met safety guidelines (retrieved January 30, 2009 from http://portal.educ.ar/noticias/educacion-y-sociedad/los-comedores-escolares-bonaer-1.php). Who is fed, what they are fed, and who is responsible for feeding students remain questions with a variety of responses.

During second wave neoliberal reform, policy discourses pay lip service to human development, or framing the student as potential capital for the national economy (Britos et al., 2003, p. 10), while a discourse of the poor as a *social risk* emerges in the haphazard administration of human development programming. Despite or in spite of a policy discourse emphasizing the importance of the well-fed student to the needs of the developing economy, the program is unevenly administered; what food is served and how the program is run remains dependent on the school community's resources. The poor become a social risk to the nation. And the risks that the poor confronted are about to increase. Responsibility for "managing" that ever-escalating risk remained in the hands of schools.

With the collapse of the Argentine economy in December 2001, the feeding program was necessary once again for the same reasons it was conceived: students needed nourishment. In 2001, the population living at risk of or living with food insecurity stood at 22.9% and that percentage grew to 40% following the crisis (Britos et al., 2003, p. 9). The percentage of students in need of nutritional supplements was the largest ever in the history of Argentina's school feeding. However, program administration at the level of provinces, districts, and municipalities was inconsistent. Provincially controlled school food programs were found to be chaotic and uneven in their ability to manage programs and budgets, and the crisis increased the demand on already functioning school cafeterias, generating a demand for more. State funding, however, was scarce and inconsistent. Some elementary and high schools had food programs; others did not. Some offered a glass of milk while others provided breakfast and/or a snack. Still others offered lunch to students.

The Buenos Aires Province, where research for this chapter was conducted, is the most populous of the twenty-three provinces with the second-largest economy next to the Autonomous City of Buenos Aires. The province's School Food Service serves a large number of students. According to figures from 2006–2007, school feeding took place in 8800 schools around the province. During the school year, 607,701 students were fed meals; 349,505 students received a glass of milk; 132,885 students received a glass of fortified milk; and 966,216 students received a complete breakfast and a snack (Chattás, 2007). What is unclear is how many adolescents receive these services or whether the numbers reflect only preschool (3–5 year-old) and lower elementary school (6–9 year-old) students fed.

Coexisting in this context, both historically and at the beginning of the twenty-first century, is poverty that leads some students to seek food assistance from their schools and prosperity that enables others to purchase snacks of choice (Sen, 1981, 1999). In the early twenty-first century, many Argentine public school students have access to food during breaks in school sessions. Cantinas continue to serve simple sandwiches, drinks, and cookies in older school buildings, and snack counters have been tucked into corners of newer ones that were not built with food preparation or distribution in mind. High school students also have the option to leave school buildings and purchase food from local businesses in commercial areas, in ground floor residential windows, or built into front yards to take advantage of a regular clientele. In school or off premises, the students' options include a limited number of healthy options such as fresh fruit, yogurt, or sandwiches of fresh bread and ham with cheese and tomato. The majority of the snacks sold are processed, including soft drinks, sweetened juices, cookies, candy, and yogurt. Students thus choose depending on their tastes and on their access to money. A rising concern is the easy accessibility of inexpensive, processed industrial foods ("junk" foods) lacking in needed nutritional content (Nestle, 2007). Poverty and the hunger that accompanies it continue to coexist with free-market choices, albeit unhealthy ones.

Argentina is a complex patchwork of nutritional ecologies ranging from the traditional manifestations of hunger, a lack of calories associated with so-called developing contexts, to malnutrition, or too many calories from poor nutritional sources reflected in contexts of modernity. School feeding is best described as fragmented, or better, chaotically localized, from its inception at the beginning of the twentieth century to the early twenty-first century. In fact it is the only characteristic of school feeding that has remained consistent throughout the course of a hundred years. At times there have been a multitude of programs administered by local authorities; at other times—brief ones—there has been one national program, though not nationally implemented. What local programs serve has varied dramatically from school to school; students might have a cup of milk—perhaps one that is fortified—or receive breakfast and a nutritional snack. Many elementary students, though, are not served at all, and some secondary students are fed, though very few. What students are fed also varies depending on what local officials purchase, prepare, and serve; some schools may serve food that meets the nutritional and caloric needs of students, others a sweet cookie (*alfajor*) and a box of sweetened juice or tea. The authors conducted their research within this context of fluctuating nutritional need, a need for nutritious school food, and a need for nutritional education.

Two School Food Problems, Same National School Food Ecology

An Educational Study

This is the story of how one public high school in the province of Buenos Aires, referred to as "Pampas" (all names are pseudonyms), actively addressed what they referred to as students' hunger—a lack of sufficient calories—and how this affected teachers' work. Hunger and school feeding were not the focus of my (Robert's) original educational study.[1] The nature of ethnography, however, opens up the possibility that a context and the persons acting within it will introduce the "professional stranger" (Agar, 1996) to a different story than one might anticipate from a review of literature or even a pilot project.

Pampas' educational professionals could not ignore the problem of hunger and instead became food advocates for their students and, later, food managers, engaging in activist work first and more service work thereafter to support their students' education. Working on behalf of hungry students, teachers also fed what I refer to as their hunger to educate Pampas' students (Robert, 2010). Like the educators, I could not ignore the problem of hunger or the school officials' long-term project of instituting a feeding program for all adolescents at the school, particularly because unemployment and limited social services continued to impact students and therefore teachers' work lives for years afterward.

The food "problem" at the center of my education-focused research began, unexpectedly, during the school's morning ritual during the 2000 school year and before the economic collapse in December 2001. A tradition emanating from the inception of mass education at the end of the nineteenth century, students at Pampas lined up in gender-segregated lines behind their teachers and in front of the school directors to hear announcements in the central patio. One morning during the ritual, a student fainted. As educators attended to the student, they were told that the student was hungry (*tenía hambre*). Administrators and teachers who were present recalled this event as a call to action. They were compelled not so much to ask why this one event happened, but instead to stop and talk with all students about life beyond school walls in the midst of a prolonged economic recession.

Educators were not blind to the ongoing economic difficulties families in the community experienced, but this event made it all seem more real, more immediate, and dire. Though the worst was yet to come, conditions in the community were not improving. Unemployment was high, so much so that when history teachers had to teach students about industrialization, many students were unfamiliar with the term for "worker" (*obrero*); why would students be familiar with the term, the educators suggested, if no one they knew had ever had regular employment (Robert, 2008). The school director, a long-time grassroots activist within the community, asked teachers and staff in the days that followed to initiate a dialogue with students in their classrooms and during breaks to discuss life beyond school walls. While calling for a dialogue, the primary responsibility of the teachers at this stage of action was to listen to the adolescents' stories. As many of the educators did not live in the neighborhood, the director urged them to listen and learn from their students so that eventually the community and faculty could work together to address the issue of hunger with as much of a shared knowledge base as possible. For many of the educators, dialogue was not a new or foreign endeavor; many chose to work at Pampas precisely because they self-identified as education activists and were compelled to work for the working poor-to-impoverished community. Pampas is located in La Matanza district, the most densely populated in Buenos Aires Province and second most densely populated area of the nation after the Autonomous City of Buenos Aires. While part of metropolitan Buenos Aires, La Matanza—and Pampas' neighborhood—did not share anything else in common with the capital city. The school, located a mere thirty miles away from the city of Buenos Aires' border, had one of the highest percentages of inhabitants living in poverty in the metropolitan region. Approximately 71% of the area's residents lived below the poverty line according to census figures from 2001 (INDEC, 2003). (In Argentina, as in many other contexts of development, the majority of poor and marginalized populations reside on the outskirts of cities, not in what in the United States is termed "inner cities.")

In classroom discussions, students echoed the initial fainting student's dilemma: they were hungry. The conversations, however, yielded a more nuanced understanding of the problem. While some adolescents came to school, others chose not to for a variety of reasons including being ashamed that their family could not feed them. Shame factored highly into reasons for not coming to school at all. In 2005 when this research was conducted, educators said that they knew which students really needed supplementary calories but would still not eat the fortified snack offered, at least not in front of classmates. Still other students explained that they needed to find quick, informal, unregulated work (*changas*) because they and their family needed the money. Adolescents, unlike elementary students, would not go to school if they were hungry.

The director also reached out to parents and community members at this time, organizing a faculty retreat that entailed home visits for the teachers and which allowed the teachers to learn by observing the students' lives beyond the school. To understand the problem, the director thought, required that educators and staff see where their students came from. Amidst tears and anger at the revelations of the humble backgrounds and, in some cases, poverty of the students, teachers came to understand their students as much more than just adolescents. The economic inequality the students experienced in their everyday lives became evident as a factor affecting any work the teachers attempted to do inside the school.

The director and faculty approach to address the problem of hunger reflects an inclusive education *for* the community and *with* the community, influenced by the critical educational approach of Paulo Freire (1973) with marginalized communities in neighboring Brazil. The teachers and director were familiar with Freire's work and with the theories of Antonio Gramsci (1971) and were able to and eager to discuss them in terms of their work at Pampas. Theories of critical educational action shaped their educational work. With student, parent, and community help, the teachers identified a problem they felt they could address collectively, at least in the short term. They resolved to address students' hunger, or lack of caloric intake (undernourishment), by instituting a snack time similar to what elementary schools already were doing.

The school community came up with a short-term plan: pool money and donations of supplies from faculty, staff, and community members to provide all students with a drink of hot tea, juice, or milk with cookies during class time. Though the high school day provides for breaks, they decided to serve the snack during an instructional period. Teachers gave up precious instructional time so that students could eat at desks in the classroom. The flow of teaching (and learning) was interrupted by snack time, another impact that hunger had on teachers' work. Teachers knew they could not continue to fund this endeavor and no budget existed to support the short-term plan beyond a budget of their own fluctuating donations. A long-term plan was needed to address, as best as their school community could do, the nutritional needs of their students.

School officials continued to collaborate with parents, students, and community members to collectively pursue a longer-term plan to address hunger at the school. Having defined the problem as undernourishment, they sought funding for food from education and social service agencies to extend the province of Buenos Aires' school feeding program, School Food Service (*Servicio Alimentario Escolar*), to Pampas' adolescents. Teachers, administrators, and community members regularly went to district and provincial government agencies to seek funding to construct a kitchen in the school's central courtyard where snacks would be prepared for all students in all school sessions by a paid staff member. This activism work was not compensated for, nor considered part of the formal teaching work that educators were contracted to fulfill. Educators volunteered—as did staff, community members, and students—to travel to the district headquarters and to the provincial capital. The collaboration was informal, with each community member donating the time, energy, and material resources they could.

The food program was not going to come to Pampas unless the teachers, directors, students, and parents fought for it. They fought, then, for an extension of the program to their school, for their students. Pampas was not included in the School Food Service program, a reflection of the chaotic and uneven nature of localized educational services that continued in the wake of educational decentralization. During fieldwork in 2005, an assistant director explained to me that the school community had to take on added responsibilities of finding social services to support students and their families and then find the families to teach them how to access the services. The statement reflects what Argentine educational researchers Dussel, Tiramonti, and Birgin (2000) described as a process through which the welfare state retracts further and further, leaving schools as the only public institution accessible to communities and resulting in teachers becoming social service providers.

Additionally, School Food Service was not going to extend to Pampas because it was a high school. Another rationale for the school community's policy approach to hunger was that hunger does not stop at elementary school. High school students from low-to-no income families needed food, too. The school-feeding model was outdated. The program was based upon a model of mass education from the turn of the previous century when an elementary education was the national goal. Throughout the twentieth century, high school attendance and completion rates had grown. According to 1999 UNESCO figures for gross enrollment rates, 85% of secondary-level-aged students attended high school (UNESCO, 2010). This number is compelling considering that mandatory schooling in 1999 did not include a high school education. The numbers reflect a desire for a secondary education among the population. Extending the food policy to include all adolescents—at least all students at Pampas—would promote educational opportunity for a sector of the population that did not have access to high school when school feeding started a century ago, catching up to the pace of early twenty-first century educational reality.

Furthermore, the school community demanded that all students, not just the most needy, be provided food assistance. Teachers and directors were concerned that students would be too ashamed to accept food distributed based on family income levels. Students told them so, and students refused the snack offered as part of the short-term plan, even students who the teachers knew needed additional food. However, the teachers' and administrators' struggle to obtain food for all Pampas' students was not just an effort to address the potential shame of hungry students. It was connected to a rights-based discourse. The students were all Argentines, they reasoned, and all deserved support from the government to pursue their education. In 1990 Argentina ratified the United Nation's Convention on the Rights of the Child. Article 24 requires that "States Parties shall ... take appropriate measures ... to combat disease and malnutrition ... through the provision of adequate nutritious foods and clean drinking water" (United Nations, 1989). The Pampas' educators were versed in and embraced the Convention (as well as other human rights-related doctrines). Beyond a right to food based on citizenship, the school professionals drew connections between the right to school food assistance and human rights.

The school's long-term feeding plan received preliminary support from district governing bodies in 2001. This victory came as a result of visits, sometimes on a weekly basis, to district education and social service offices and periodic visits to equivalent provincial governing bodies hours away. Funding for the construction of a kitchen, a staff member to prepare and distribute food, and delivery of food, came first. However, funding was not sufficient to provide the snack to Pampas' students universally. The school community refused the funding unless all students would receive the assistance. Months of lobbying and uncertainty continued until sustainable funding of a snack for all students was won. By the 2004 school year, the universal school-feeding program was in place. This activist phase of extra work for the teachers and administrators related to school feeding came to a close.

In July 2003 and 2005 when this research was conducted, administrators and teachers were satisfied that they had addressed to the best of their ability the feeding problem that was plaguing some of their students. Their snack program had extended the Argentine tradition of school feeding to include high school students in an economically needy community in an ongoing economic recession. The problem of hunger, or undernourishment, was addressed, and funding—for the time being—was secured. Teachers returned to the work of educating the nation's poorer students for a weak economy with little promise of employment. However, due to the nature of the program that was funded, teachers' classroom work was compromised by a daily interruption for snack distribution and consumption. Instruction came to a standstill every day while a staff member walked around the class handing out a traditional cookie (*alfajor*), drink boxes (milk or juice), or a sandwich, depending on the food delivery.

In the previous pages, I have shared the story of how the province of Buenos Aires' School Food Service was brought to Pampas. Teachers, in conjunction with school directors, staff, students, parents, and community members, defined the school-feeding problem as a lack of sufficient calories. High school teachers' work morphed through the process of defining the problem, identifying and instituting a short-term plan, obtaining funding for the long-term plan, and, last, institution of the program. First, their work involved caring for the immediate needs of their high school students—from attending to a fainted student to identifying students' unmet basic needs in an ongoing economic recession. *Profesoras/es*, as high school teachers in Argentina are called, differentiated themselves from their elementary, or *maestra/o*, counterparts by a lack of caring involved in their job description. While elementary educators were deemed caregivers as part and parcel of their educational work, secondary subject-specific educators viewed their work as imparting academic knowledge. Though many Pampas' educators embraced critical educational practices, they viewed their work to be primarily about teaching students content. This unqualified care-giving work continued through the short-term feeding plan until the long-term funding was obtained.

Second, Pampas teachers worked as activists as they sought funding and resources for a school food program. Though many educators considered themselves critical educators and activists because of their work with marginalized populations, the activism involved in obtaining school feeding took place outside the classroom and even the school. It required time, energy, and material and human resources as the teachers traveled to different governmental agencies to press for a Pampas' food program. This was the most intensive work educators took on and they took it on beyond what is considered their workplace. They volunteered to extend their work boundaries. However, the nature of government programming in a decentralized Argentina required them to do so if they were to be able to do their work educating in the classroom. Teachers cannot teach hungry students well, and they cannot teach students who do not come to school because they are hungry.

The last phase of the project seemed to have brought teachers' work back to the classroom, but at times the classroom became a cafeteria. Instruction was interrupted for snack time. Teachers needed to find a way to blend their hunger for educating working poor students with their students' hunger for food. Teachers must now balance the work of educating with that of food worker or food manager and social worker. Teachers' work has been perhaps both eased by the food program and intensified, or at least the number of roles to be fulfilled has multiplied.

I retold this story to Irina Kovalskys, a medical doctor with a specialization in pediatric nutritional health. She was surprised by the term hunger (*hambre*) that I borrowed from the educators. From her perspective in the medical community, Metropolitan Buenos Aires, where Pampas is located, is a school food ecology riddled with *mal*nutrition rather than *under*nutrition. Her own research of schools in the metropolitan region revealed a different problem: malnutrition and obesity, from that identified and addressed by the educational community.

Two Public Health Studies

In the early 1900s immigrant families populated Argentina, bringing with them hunger caused by war in Europe. Now, in the early years of the new millennium, Argentina is defined by globalization and the nutritional transition phenomenon (Popkin, 2001): a significant decrease in infant mortality, a decrease in acute malnutrition indicators such as anemia or children who are small for their age, and an increase in obesity and chronic diseases like diabetes and heart disease. Populations experiencing this phenomenon continue to struggle with undernutrition, but higher percentages continue to become overweight and obese (Kain, Corvalán, & Uauy, 2010). This is attributed to changes in the availability of food, both in terms of quantity and quality. Food culture and the industrialization of food also have changed. From a nutritional point of view, the studies described in this chapter, as well as other studies in Latin America (Uauy, Albala, & Kain, 2001), are consistent in showing a tendency toward increased consumption of baked goods—often times highly processed industrial food products, not homemade—high in fat (such as sweet and salty cookies and pasties) and simple sugars. The habit of drinking of water as a main source of hydration is replaced with the consumption of sweetened beverages, added to the decreased consumption of fruits and vegetables. The social, demographic, and economic transformations suffered by developing countries like Argentina in recent decades resemble this epidemiologic process of nutritional transition phenomenon.

Cyclical economic crises also have exposed Argentines to the risks of hunger and food insecurity. Economic crises also have affected public health workers exposed on a daily basis to patients and clients facing hunger or food insecurity. Both adolescents and public health workers were immersed in the aforementioned epidemiological context of nutritional transition when at the end of 2001 Argentina faced one of the worst economic crises in recent decades. The exact impact of the crisis on the nutrition of adolescents has not been studied extensively, partly because epidemiological health data is insufficient and because, particularly for the adolescent population, it is difficult to obtain an adequate sample. For these reasons, public health data on this age group's state of nutrition is irregular and imprecise, especially for this time period.

Another reason that the data for this group is limited is that historically the majority of public health interventions in Argentina have focused on the period of greatest vulnerability: the first two years of life. In fact one of the largest and the oldest public health programs, the Maternal Infant Program (*Programa Materno-Infantil*, PMI), served pregnant women and their infants. It has evolved into the Maternal and Infant Nutrition Program (*El Programa Materno Infantil y Nutricional*, PROMIN) and programming has evolved as a result of food insecurity due to economic crises. In light of growing food insecurity and related health concerns, the program also expanded to promote healthy lifestyles through nutritional counseling, expanding public health interventions into schools, once again, with a focus on nutritional education. Currently, nutritional guides are being produced for the preschool and elementary aged population (Ministerio de Salud, 2006).

While the new phenomenon of nutritional transition takes place simultaneously during an economic crisis, children move through adolescence, the period from puberty to adulthood. Puberty, as a physiological period, is the beginning of a dynamic process of growth and psychological and social development that involves adjustments and adaptations with the environment. And it is a period of high nutritional vulnerability (Walker & Watkins, 1997). There is an increase in lean mass, seen in the increase of the size of cells, tissues, and organs. Ideally all such biological changes would be accompanied by nutritional balance, optimizing healthy growth and development. Excess calories and nutrient deficiency during this period alter this balance and produce excess fatty tissue in the first case, or malnutrition in the second case, which, depending on their degree, can compromise physical growth and puberty development.

In 2002, the public health community's concern centered on addressing hidden deficiencies in foods through laws requiring the addition of thiamine, riboflavin, niacin, folic acid, and iron in flour (Congreso Argentino, 2002). This is complemented by fortification of milk distributed through social development and health programs (Duran, 2007). In 2003 as a way to build data on adolescents, specifically, in the years following the crisis, I (Kovalskys) analyzed nutritional variables of pregnant adolescents (N=125) under 18 years of age at the beginning of pregnancy, or before 12 weeks of gestation (Kovalskys,

2006), a population that I have access to through my clinical work attending patients in public and private institutions in metropolitan Buenos Aires. Even though the adolescents in the study did not attend Pampas, their socio-economic status is similar to that of Pampas' students as is the vulnerability of being an adolescent at the particular moment the study was conducted; 50% of participants attended public schools, though none attended Pampas.

Taking into account the social and economic scenario of Argentina at the time, the first hypothesis of the research was that adolescents would have limited access to food and would show low weight as a sign of clinical nutritional deficit and/or anemia as a hidden nutritional sign. However, the findings were not as expected: only 2% of the population was underweight, and 20% was overweight. The prevalence of anemia was 18% with no direct link between both weight variables. The initial analysis of the intakes showed that 17% exceeded the daily calorie requirement (more than 2,800 kcal) and approximately half the population reported an insufficient intake of meat, leading to a low intake of iron (less than 12 mg/day). The study showed that two years after the acute economic crisis and still suffering the economic consequences, the participants met caloric needs with or without government or community assistance. Nutritional deficit was reflected more in the low nutritional value of the calories obtained rather than in a lack of daily calories, leading to overweight or obese and iron deficient young women.

The International Life Sciences Institute-Argentina designed a second study in metropolitan Buenos Aires to determine the prevalence of overweight and obese primary school children (N=1,588) and to ascertain the main associated factors that could be linked to the environmental context. Conducted from 2005 to 2006, it was the first epidemiological investigation to determine the nutritional status of boys and girls right before the beginning of adolescence (10-to-11-year-olds). All of the children attended public schools. The objective of the study was to assess the prevalence of childhood overweight and obesity and its associated factors including nutritional variables such as height for age and other signs of chronic malnutrition (stunting). Prior to the study, the absence of data regarding overweight and obese children made it difficult to identify effective and accurate intervention strategies. The main findings were that prevalence rates of children who are overweight, including those who are obese, are comparable to those of countries that consider this to be a problem of epidemic proportions, affecting a high percentage of the studied population or a range of 27.8% to 35.5% depending on the reference used. Mean thinness frequency was lower than the expected frequencies of 3% according to the World Health Organization or 5% for Centers for Disease

Control, except for 11-year-old girls (Kovalskys, Rausch Herscovici, & De Gregorio, in press).

When analyzing eating behaviors and food consumption, the findings showed that only 2% of the children in the study consumed an adequate amount of vegetables according to the recommended allowances for the Argentine population (equivalent to two times per day) (Asociación de Dietistas y Nutricionistas Dietistas, 2001). Twelve percent consumed vegetables once a day and the remaining 86% consumed vegetables less than once a day. The consumption of fruit in this population was also deficient (Kovalskys, Holway, Ugalde, & De Gregorio, 2007). Participants' 24-hour intake logs also showed an average caloric intake of 2,316 kcal per day, 16% more than the recommended 2,000 kcal. The average protein intake surpassed the recommendations, and the average fat consumption was 84 g per day, compared to the maximum recommendation of 77 g per day (Institute of Medicine, 2002). The availability and consumption of unhealthy food—in terms of calories and lack of nutritional value—also was observed in schools.

One example of a public health sector intervention to address the issues of overweight and obese students and unhealthy food consumption in schools is the Healthy Canteen Program (*Cantinas Saludables*) in the city of Rosario since 2005 (Secretaria de Salud, 2005). The program focuses not only on increasing healthy options at schools' canteens but also on promoting physical activity and increasing students' knowledge about healthy eating. An evaluation of that program found that the overweight and obesity rates found in the studied population warrant the intensification of prevention programs because school canteens are an important provider of junk food (Kovalskys, Rausch Herscovici, & De Gregorio, 2008). Obesity prevention programs should clearly target and intensify actions aimed at improving/changing the food options provided by school canteens.

The studies of overweight and obese public school students and of school food consumption provide evidence of a high prevalence of overweight, anemic adolescent females and overweight and obese students about to enter adolescence. The research highlights the need for public policies that are not limited to pre-adolescent children and women of child-bearing age, and it justifies public policies designed to prevent children from becoming overweight and to stimulate healthy nutrition from an early age onward. It also points to the need to complement school nutritional guides with more intervention programs in schools focusing on: the improvement of eating habits; outreach programs to promote and implement policies that regulate the nutritional quality of foods offered at school; and the increase and promotion of regular physical activity.

Multiple Definitions of Hunger Coexist

Hunger can be associated with the nutritional needs of students, but it also can be associated with socio-cultural, political-economic, and psychosocial context. The Pampas' student may have indeed passed out as a result of nutritional needs. Whether the students' nutritional need would be better defined as a need for food (undernourishment or a need for more calories) or for iron-rich food (malnutrition associated with lack of nutrients particularly iron) is unclear. The purpose of this chapter was not to define the particular student's problem. The incident brought the two authors together to better understand the school feeding problem in the Argentine context simply defined in everyday terms as hunger.

Hunger has many definitions in everyday language. The word "hunger" signifies, in its most basic form, a need for calories, for nutrients, and even for the act of eating and all that the act entails socially, politically, economically, spiritually, and psychologically (Mintz & Du Bois, 2002). Our studies of food in Argentina's schools show how students' needs are understood differently by different professionals all concerned with students' well-being. Furthermore, our studies show how students' nutritional needs are understood differently by those different professionals. How school and public health professionals understand those needs varies based on epistemological approaches to studying and defining the problem of hunger. Teachers explored the problem from the perspective of their lived, or socio-cultural, experiences and those of their students and their families. The problem of hunger was linked to the broader political-economic context in which those experiences emerged, but most importantly were linked back to the educational project that put them in the position to study their students' hunger problem in the first place. Public health professionals explored students' physiological qualities and their eating habits inside and outside of school. They sought to quantify the adage, "you are what you eat," in terms that related back to broader public health projects to divert increasing rates of poor health associated with obesity, for example.

The lack of (nutritious) food and concern for overweight and obese children are not oppositional phenomena, but are overlapping phenomena in nations around the world. Our findings suggest that both groups' pressure to amend Argentina's feeding programs address ecological concerns in the highly decentralized school food program. In the localized ecology of school food, educators pressed for the current program to be extended to feed their high school students only. Their efforts were localized due to the scope and immediacy of the problem. Their efforts also reflect school communities' ability to access state services in a decentralized social service context. Government officials are not going to come to schools with funding, even funding deemed crucial to the health and education of the nation's students. Communities must now go find services, fight for them, and adapt them to accommodate students' unmet needs. The school professionals' push for food did not include an explicit push for highly nutritious food. Getting calories for their students was the primary concern; if it was nutritious, it was a bonus. Conversely, but complementing the drive to get food to students, public health efforts to measure and address malnutrition have focused on increasing the nutritional value of the foods consumed and increase the nutritional education students receive along with the healthier fortification. The public health professionals may have access to government institutions, but they too must seek out spaces where their concerns are a priority and fit with the current institutional structure. Both groups, however, were found to be tapping into different institutions, education and public health, showing that many actors are working to intervene in school feeding but through different institutional channels.

Policy Implications

To consider hunger—as we do—as a multi-definitional problem encompassing both lack of calories and an excess of nutritionally "empty" calories has a number of implications for Argentine social and educational policy. We believe, first, that any lack of food should be fulfilled with nutritional food. All students would be better served through school food programs that provide nutrient-rich calories that tap into the equally rich production of fresh, healthy foods locally from fruits and vegetables to whole grains to fresh non-sweetened milk.

> The food that any program distributes (or promotes for consumption) should be appropriate for addressing nutritional problems of the population. Also it is important that food is related to the food culture [where it is served] and is not complicated to prepare, that the distribution is systematic, without interruptions, and accompanied by a nutritional education component. (Britos et al., 2003)

In addition to being healthy, then, the food procured by or delivered to schools should be easy to prepare, not requiring intensive labor on the part of educators, or when necessary funds should cover the additional labor needed to prepare fresh meals.

Second, and related to the first suggestion and quote above, nutritional education should accompany a healthy food program whether designed as a separate curricular unit or integrated into other subject areas. School feeding continues to be framed as a means to "manage the poor" or is justified as a way to offset social risks involved in not "helping" poor families (Leguizamón, 2005, p. 10). School food policy perhaps should and could be linked to broader discourses of human rights, and children's rights specifically, which are already the framework for teaching civics within Argentina's educational institutions and for providing public health services. A right to nutritious food can be foregrounded, especially in the context of ongoing economic hardship within the space of existing curricular areas.

Schools are a unique space to teach about health and can play an important role in preventive health care. Health education can teach students healthy eating habits and to take care of the body through healthy food "curriculum" even in the space of the cafeteria (see DeLeon, this volume) and, of course, through classroom curriculum. It is essential to train educators and health agents, however, in food and adolescent nutrition too. This can start with conversations regarding the multiple definitions of hunger that define the Argentine ecology from undernourishment to malnutrition.

Third, we believe that nutritional programs in Argentina should extend to secondary education. Argentina's rate of high school attendance and new education law that mandates compulsory education through secondary education (Ley de Educación Nacional, 2006) are clear rationales for why the extension of this policy should serve a broad population and, in the process, strengthen current educational mandates and opportunities. Again, linking school feeding programs to a rights-based discourse is a strong rationale for such an extension. Additionally, feeding more students healthy food should be seen as offsetting concerns for increased cost. As the public health studies illustrated, overweight and obese students should be viewed as a long-term societal and political-economic concern.

Finally, the conversations that led to this chapter reflect the need for more dynamic social policy analyses that move beyond techno-managerial models (Hintze, 2005). Such approaches hold more potential for uncovering complex understandings of youths' needs. This is particularly true of national ecologies such as Argentina, where definitions of the school food problem and how to address the defined problem vary dramatically depending on the actors involved and the location within the ecology of those actors.

We do not call for education professionals to become public health professionals or vice versa. Both groups of actors bring to the policy analysis vital knowledge for improving school food programs. Teachers' work in conditions of poverty is taxed enough. Instead, we call for teachers and public health professionals to move toward relationships characterized by cooperation, explicitly working together to accomplish mutual goals. Though the definitions of the school feeding problem seemingly diverged, both groups of actors' concerns converged on adolescents' nutritional and educational needs. While both may continue to work on a day-to-day basis within bound environments, much overlap exists in their work in schools. It is clear that educators have agency and are willing to act on behalf of their students. And it is clear that public health professionals are eager to help schools craft a healthy meal program and nutritional education workshops to accompany it. Cooperative work toward policy change means that both groups come to the decision-making table with more input, more power, and representing more niches within the policy ecology. Relationships must be developed that link these diverse actors in the cause of defining collaboratively how school feeding policy should be reformed. This chapter, in other words, is a continued call for more inclusive and participatory policymaking.

Conclusion

Conversations among different actors with agency in the policy ecology offer one potential means through which to better define school feeding problems in Argentina and—we would suggest—many other nations. A shift in relationships from quasi-symbiotic to cooperative opens up definitions of the problem that school-feeding programs can address. Our conversations across epistemological paradigms revealed the complexity inherent in any attempt to collaborate within a complex policy ecology. We had to learn from each other about the school food ecology, define and redefine the parameters of the problem. The approach we have taken to rethinking Argentina's school food program offers an alternative construction of the school food problem and how to break through an ecology suffering from entropy. While efficiency of distribution should always be a concern, such discourses are buttressed by beliefs that recipients are "milking the [social welfare] system" or, worse, not entitled to basic needs or to full citizenship rights. Yet the recipients of food programs are citizens, or future citizens, whose education should be paramount to building and maintaining a participatory democracy. Providing nutritious food so that students can focus on learning, and providing knowledge of the important choices to be made each and every day toward supporting one's health, should be the goals of school food programs in a democratic society. School feeding must be framed in terms of rights discourses and citizenship definitions. When it is not, we call on educators and public health professionals to collaborate or at least engage in a collaborative dialogue to learn from each other and, hopefully, to bring a healthier school food program to "fruition." To start, try defining the "problem."

Note

1 Sarah A. Robert's research was generously supported by The Tinker-Nave Travel Grant for Pre-Dissertation Fieldwork from the University of Wisconsin-Madison's Caribbean, Iberian, and Latin American Studies Program, and the Fulbright-Hays Doctoral Dissertation Research Abroad Fellowship.

References

Agar, M. (1996). *The professional stranger: An informal introduction to ethnography* (2nd ed.). San Diego: Academic Press.

Asociación de Dietistas y Nutricionistas Dietistas. (2001). *Guías Alimentarias para la Población Argentina*. Retrieved from http://www.fmed.uba.ar/depto/edunutri/gapa.htm

Billorou, M. J. (2008). El surgimiento de los comedores escolares en la Pampa en crisis. *Quinto Sol* (12), 175–200.

Britos, S. A. (1995). Argentina: Reformulación de programas de comedores escolares. *Aberto, 15*(67), 73–81.

Britos, S. A., O'Donnell, A., Ugalde, V., & Clacheo, R. (2003). *Programas Alimentarios en Argentina*. Buenos Aires: Centro de estudios sobre nutrición infantil (CESNI).

Buamden, S., Graciano, A., Manzano, G., & Zummer, E. (2010). Proyecto "Encuesta a los servicios alimentarios de comedores escolares estatales" (PESCE): Alcance de las metas nutricionales de las prestaciones alimentarias de los comedores escolares de Gran Buenos Aires, Argentina. *DIAETA, 28*(130), 21–30.

Chattás, J. (Producer). (2007, October 10, 2010). Los comedores escolares bonaerenses no se toman vacaciones [Press release]. Retrieved from http://portal.educ.ar/noticias/educacion-y-sociedad/los-comedores-escolares-bonaer-1.php.

Drewnowski, A (2004). Poverty and obesity: the role of energy density and energy costs. *American Journal of Clinical Nutrition, 79*, 6–16.

Duran, P. (2007). Anemia por deficiencia de hierro: estrategias disponibles y controversias por resolver. *Archivo Argentina Pediatría, 105*(6), 488–490.

Dussel, I., Tiramonti, G., & Birgin, A. (2000). Towards a new cartography of curriculum reform: Reflections on educational decentralization in Argentina. *Journal of Curriculum Studies, 32*(4), 537–559.

Ewig, C. (2010). *Second wave neoliberalism: Gender, race, and health sector reform in Peru*. University Park, PA: Pennsylvania State University Press.

Freire, P. (1973). *Education for critical consciousness* (1st American ed.). New York: Seabury Press.

Gramsci, A. (1971). *Selections from the prison notebooks of Antonio Gramsci*. New York: International Publishers.

Hintze, S. (2005). La evaluación de políticas sociales en la Argentina: reflexiones sobre el conflicto y la participación. In L. Andrenacci (Ed.), *Problemas de política social en la Argentina contemporánea* (pp. 157–180). Buenos Aires: Prometeo.

INDEC. (2003). *Incidencia de pobreza y de la indigencia en el Gran Buenos Aires.* Buenos Aires: INDEC.

Institute of Medicine. (2002). *Dietary reference intakes for energy, carbohydrate, fiber, fat, fatty acids, cholesterol, protein and amino acids, part I.* Washington, DC: National Academy Sciences.

Kain, J., Corvalán, C., & Uauy, R. (2010). Developing country perspectives on obesity prevention policies and practices. In E. Waters, B. Swinburn, J. Seidell & R. Uauy (Eds.), *Preventing childhood obesity: Evidence, policy, and practice* (pp. 283–291). Oxford, UK: Wiley-Blackwell/BMJI Books.

Kovalskys, I. (2006). Impacto del estado nutricional de la embarazada adolescente al comienzo de la gestación en la salud de la adolescente, del recién nacido y perinatal. En *Becas Ramón Carrillo-Arturo Oñativia* (pp. 10–74). Retrieved from http://www.saludinvestiga.org.ar/libros_comision.asp. Buenos Aires: Ministerio de Salud y Ambiente.

Kovalskys, I., Holway, F., Ugalde, V., & De Gregorio, M. J. (2007). *Análisis de los Factores vinculados a sobrepeso y obesidad en niños de 10 y 11 años que asisten a escuelas públicas en el area metropolitana de Buenos Aires.* ILSI Report Series.

Kovalskys, I., Rausch Herscovici, C., & De Gregorio, M.J. (2008). *Evaluation of a school-based obesity prevention program in Rosario, Argentina. Final Report to ILSI Research Foundation.* Buenos Aires: ILSI-Argentina.

Kovalskys, I., Rausch Herscovici, C., & De Gregorio, M.J. (in press). Nutritional status of school-aged children of Buenos Aires, Argentina: Data using three references. *Journal of Public Health.*

Leguizamón, S. A. (2005). La invencíon del desarrollo social en la Argentina: Historia de "opciones preferenciales por los pobres". In L. Andrenacci (Ed.), *Problemas de política social en la Argentina contemporánea* (pp. 81–124). Buenos Aires: Prometeo.

Ley de Educación Nacional, 26.206. (2006). Buenos Aires: Congreso de la Nación Argentina.

Lionetti, L. (2007). *La misión política de la escuela pública: educar al ciudadano de la república (1870–1916).* Buenos Aires: Miño y Dávila.

Mazza, C.S., Ozuna, B., Krochik, A.G., & Araujo, MB. (2005). Prevalence of type 2 diabetes mellitus and impaired glucose tolerance in obese Argentinean children and adolescents. *Journal of Pediatric Endocrinology and Metabolism, 18*(5), 491–498.

Ministerio de Salud. (2006). *Guias alimentarias para la población infantil. Orientaciones para padres y cuidadores.* Buenos Aires. Retrieved from: http://www.msal.gov.ar/htm/site/promin/UCMISALUD/publicaciones/pdf/PDF_Padres_baja.pdf

Ministerio de Salud. (2007). *Encuesta Nacional de Nutrición y Salud (ENNyS). Documento de Resultados.* Buenos Aires.

Mintz, S. W., & Du Bois, C. M. (2002). The anthropology of food and eating. *Annual Review of Anthropology, 31,* 99–119.

Nari, M. (2004). *Políticas de materinidad y maternalismo político: Buenos Aires, 1890–1940* (Maternity politics and political maternity: Buenos Aires, 1890–1940). Buenos Aires: Biblos.

Nestle, M. (2007). *Food politics* (Revised and expanded ed.). Berkeley: University of California Press.

Peña, M. & Bacallao, J. (2000). *Obesidad en la pobreza.* Publicación Científica 576. Washington, DC: Organización Panamericana de la Salud.

Popkin, B.M. (2001). The nutrition transition and obesity in the developing world. *Journal of Nutrition, 131*(3), 871S–873S.

Puiggrós, A. (1990). *Sujetos, disciplina y curriculum en los orígenes del sistema educativo argentino* (Subjects, discipline, and curriculum in the origins of the Argentine educational system) (Vol. I). Buenos Aires: Editorial Galerna.

Robert, S. A. (2008). *Gender, Education Reform, and Teachers' Labor in Argentina.* Ph.D. Dissertation, University of Wisconsin-Madison.

Robert, S. A. (2010). A Hunger for Education: High school feeding programs, teachers' work, and education politics in Argentina. Paper presented at the AERA Annual Meeting, Denver, CO.

Secretaria de Salud.(2005). *Municipalidad de Rosario. Programa cantina saludable.* Retrieved from http://www.rosario.gov.ar/sitio/salud/alimentacion_saludable_cantinas.jsp.

Sen, A. (1981). *Poverty and famines: An essay on entitlement and deprivation.* Oxford, Clarendon: Oxford University Press.

Sen, A. (1999). *Development as freedom.* Oxford: Oxford University Press.

Subcomisión de Epidemiología y Comité de Nutrición. (2005). Consenso sobre factores de riesgo de enfermedad cardiovascular en pediatría. *Archivo Argentina de Pediatría, 103*(3), 262–281.

Transfer Law (*La Ley de la Transferencia*), 24.049, Congreso de la Nación Argentina (1992) (enacted).

Uauy, R., Albala, C., & Kain, J. (2001). Obesity trends in Latin America: transiting from under- to overweight. *Journal of Nutrition, 131*(3), 893S–899S.

UNESCO. (2010). *Education in Argentina.* Retrieved October 30, 2010, from http://stats.uis.unesco.org/unesco/TableViewer/document.aspx?ReportId=121&IF_Language=eng&BR_Country=320&BR_Region=40520.

United Nations. (1989). *Convention on the Rights of the Child.* Retrieved from http://www2.ohchr.org/english/law/crc.htm

Walker, W.A., & Watkins, J.B. (1997). *Nutrition in pediatrics: Basic science and clinical applications.* London: People's Medical Publishing.

Weaver-Hightower, M. B. (2008). An ecology metaphor for educational policy analysis: A call to complexity. *Educational Researcher, 37*(3), 153–167.

•CHAPTER FIVE•

Free for All, Organic School Lunch Programs in South Korea

Mi Ok Kang

There is little doubt that the school should care for students' health and well-being as well as their academic achievement. For this reason, carefully considering how the school can best support socioeconomically marginalized students through sound school lunch systems has been a core issue for researchers and policymakers worldwide (Levine, 2008; Morgan & Sonnino, 2008; Poppendieck, 2010). The scope and current status of nations' school food programs have been quite varied based on their economic, social, ecological, and political environments. Policymakers and researchers in many nations have developed school food policies depending on their domestic conditions, as well as by borrowing or copying ideas from other nations (Hall, 1986; Ball, 1994, 1998). Therefore, nearly every local policy is in some way connected to the policies of a globalized world. This is certainly true for the new school lunch systems in South Korea.

This chapter traces the policy development for organic, universally free lunch systems for children in South Korea. South Korean society recently increased opportunities for developing sustainable school lunch systems in anticipation of the June 2010 election, which was the first direct election nationwide for Superintendents of Education and for Local Boards of Education in South Korean history. Candidates' election pledges were of interest to the public, promising changes across a host of social, ideological, and political environments. Progressive groups won the election in most areas nationwide. The victory occurred in large part because the progressives' major agenda, "free for all, organic school lunch," provided the public with a decisive chance to realize the importance of social welfare programs and policies that have been shrunken under the current conservative government since 2008.

This paper (a) provides a brief history of school lunch systems in South Korea since 1953, and (b) analyzes the discourses and texts regarding school lunch systems generated from 1992 to 2010, specifically focusing on the differentiated discourses and practices along the socio-cultural, political, and ideological lines of progressives and conservatives. Examining the discursive developmental process for sustainable school lunch systems provides ideas for policymakers and researchers across education. Specifically, the South Korean case illustrates (a) a public-driven educational policy agenda which is leading to social change, (b) a model for challenging neoliberal forces that encourage competition rather than collaboration, and (c) a sound example of how universal free lunch can be justified within the context of limited state resources and a retrenchment of government-provided social services. I will use an ecological approach (see Weaver-Hightower, 2008), contributing critical analysis to the complexity of policymaking processes in South Korea.

Approach to Analysis

Understanding the complexity of policymaking processes is crucial both to educators and policymakers who strive to create better educational environments. Although the complexity of social, cultural, and political environments around the implementation of various school lunch programs has been explored in many nations including the United States, the United Kingdom, and Italy (Levine, 2008; Morgan & Sonnino, 2008; Poppendieck, 2010), a procedural approach for considering disparate influences on policy has been missing from the research. Over the past two decades, the "ecology" metaphor has become popularized by many researchers of policy (Goodlad, 1987; Baker & Richards, 2004; Weaver-Hightower, 2008), and for good reason: the "ecology" metaphor is critical for educational policy analyses because an ecology expands the space by which researchers can analyze and discuss the inherently *bumpy* process of educational policy creation and implementation (Weaver-Hightower, 2008). Ecological analyses of often-rocky policy terrains give researchers and policy-makers ideas for creating, developing, and managing sustainable policies. Building on these ecological perspectives, this chapter charts a history of the development of a sustainable school lunch ecology in South Korea (1953-present), specifically focusing on (a) how political decisions regarding school lunch programs have shaped the system, and (b) how various agents and agencies have clashed, generating competing discourses and texts in the race for political victory.

To accomplish this, I analyzed 101 newspaper articles obtained online under the search terms of "organic school lunch," "environment-friendly school lunch," and "free for all school lunch," all keywords found frequently in the debate. For this chapter, I analyzed articles from progressive newspapers (*Hangyorae* and *Gyunghyang*) and conservative newspapers (*Chosun*, *JungAng*, *Dong-A*, *Saegae*, and *GukminDaily*). Most of the articles searched for this chapter were published between 2009 and 2010, while a quarter of them were generated in 1992, 1997, 2002, and 2007, corresponding with South Korean presidential elections. The media data analysis is triangulated with a review of legal documents such as the *School Lunch Law* (MOE, 1981, 1993; MEHR, 1997, 2006), the *Hazard Analysis and Critical Control Point* (MHW, 2000b), and the regulations of *General Agreement on Tariffs and Trade* (GATT; WTO, 1994). Finally, I examined the official websites of major civic groups (People's Solidarity for Participatory Democracy, 1994; The School Lunch Network Nationwide, 2002; The National Movement Camp for Safe School Lunch, 2003) for the most current information regarding school lunch policies. Through these sources, the various agents and agencies working for (or against) free school lunch programs were examined in line with their agendas.

From Zero: Foundation and Expansion of School Lunch Systems, 1953–2002

At the end of the Korean War, Korea lacked resources for, and models of, school lunch programs. The Canadian government launched the South Korean lunch system as a relief activity in 1953. Providing young students with food was an urgent need in impoverished post-bellum conditions, and the schools played an active role in students' survival by distributing powdered milk. Later on, foreign support expanded to include other relief agencies such as UNICEF, CARE, and US-AID. Intervention through these agencies was the only way to save students from hunger. Dried skim milk, corn powder, vegetable oil, and flour were provided to elementary students until foreign aid was officially terminated in 1973 (K. Choi, 2006).

There was scant governmental support for school lunch programs, so people in some rural areas began to grow greens and raise poultry for school lunches in their communities. Urban schools served bread and milk only to hungry students or began to equip schools with food service equipment to prepare lunches for children in need. The government abolished the urban bread distribution program in 1977 when a mass outbreak of food poisoning took the life of one student. Consequently, there was a push for the government to fund special facilities for food service delivery, but only one in ten elementary schools served school lunches by the late 1980s.

The anemic percentage of lunch programs can be attributed to flux in governmental priorities. The Korean government paid much attention to the safety and quality of school lunches in the text of the *School Lunch Law* of 1981

(K. Choi, 2006), but the agenda under a harsh military regime led by President Doohwan Jeon (1980–1987) focused on economic development rather than on social welfare, resulting in poor funding for school lunch systems.

There were viable counter-hegemonic movements for bringing social justice and democracy to society, but strong governmental control kept movements' activities largely underground. Discourses on social welfare were often prohibited, and the authoritarian government repeatedly used the phrase "welfare state" pejoratively; thus, promoting school lunch systems as a social good was almost impossible. Loosened governmental control during democratization in the late 1980s offered the potential for much better social welfare programs, and it was not surprising that beginning comprehensive school lunch systems at every elementary school was one of the most pressing items on the national agenda in the early 1990s.

In 1992, only 11.3% of elementary schools had lunch programs, serving 17.4% of all elementary students (Ahn, 1992). As suggested, however, growth was to come, and structural reorganization was a crucial part. In a nation-state where "education fever" (Lee, 2005) was spreading as a quality education was assumed to be the best way to ascend the socio-economic ladder, and in a culture where a meal is assumed to be a set combination of rice, soup, kimchi, vegetable, and fish or meat (quite hard for mothers to pack!), one of the most attractive pledges of the 1992 presidential election was setting up school lunch systems nationwide. All presidential candidates announced plans for expanding school lunch programs to all elementary schools.

President Yeongsam Kim (1993–1997) tried to fulfill his pledge after he won the election by requiring local community leaders to pay for school lunch facilities instead of increasing the education budget, which encountered serious criticism among the public. Kim announced a revision of the *School Lunch Law*, requesting that every regional community organize a Committee for Propelling School Lunch to do fundraising from a variety of sources (e.g., business entrepreneurs, regional socio-political leaderships, alumni members, parents). This policy triggered serious debate and criticisms with many people arguing that lunches should be a national, rather than a regional, fiscal responsibility (Sim, September 15, 1992). However, the government actually cut all education budgets for school lunch systems in 1994 without any reasonable explanation aside from tight governmental budgets, and it announced a steep rise in lunch monies expected from families. The public became upset, petitioning against increased out-of-pocket lunch expenses and hosting several open forums against the new policy (Ji, 1994). Likewise, parent associations were organized nationwide, and they conducted campaigns for school lunch programs (Yoo, April 23, 1995).

Those efforts, however, did not fully affect real policymaking processes until the next president, Daejung Kim (1998–2002), adopted a compulsory elementary school lunch system as one of his election pledges and promoted it as soon as he won the election. After setting up school lunch systems with reasonable funding for every elementary school in 1998, Kim endorsed an enforcement ordinance for school lunch expansion in every high school. This ordinance made available governmental funds for equipping schools with kitchens and gave private suppliers permission either to cook meals at the schools or to deliver meals to the schools (K. Choi, 2006). As the high school lunch policy was embraced successfully in 1999, the government then drew up the budget for middle school lunch programs by 2002. Government data in December 2004 showed that 99.9% of elementary, 97.8% of middle, and 98.7% of high schools provided school lunch programs and that 94.3% of elementary, 94.7% of middle, and 85.8% of high school students received benefits from those programs (H. Choi, December 21, 2006), a sizeable increase in the magnitude of school lunch programs. Considering that the South Korean economy began experiencing a harsh foreign currency crisis in 1997 that spurred budgetary retrenchment, implementing national school lunch programs that reached the vast majority of students was quite remarkable.

There were certainly tensions and obstacles, however. Conservative groups challenged the Kim government, arguing that plans for *quality* school lunch programs deserved priority over expanding the quantities of children served, and they argued that it was too early for the government to promote high school lunch programs in such serious economic circumstances. However, the government sped up the establishment of school lunch programs in every K–12 school, an initiative that received strong public support in the wake of decreasing family resources (Lee & Hong, April 4, 1998).

The public now had space to address three major issues still left behind: (a) how to guarantee quality lunches with locally grown, organic, environmentally friendly foods, (b) how to transition from private-entrusted lunch systems to school-managed ones that ensure quality and safety, and (c) how to better serve students in poverty without hurting their self-esteem. The next section examines the complex processes by which social agents and agencies have tried to resolve problems of where school lunch foods should come from and how they should be funded and distributed.

An Era of Debates, 2003–2009: What Should Be Done Next?

Aligned with the concerns addressed in previous years, three agenda items were set forth by the School Lunch Network Nationwide, the Association of Civic Organizations in Solidarity, and the Headquarters of National

Movements for Enacting/Revising the School Lunch Law and Ordinances in solidarity with about 650 civic groups, progressive and radical political parties, and labor unions. The School Lunch Network Nationwide specifically proposed that school lunches: (a) be school-managed, (b) be available free to all students, and (c) use organically grown local foods. This section examines the discourses and texts of this agenda, unpacking the ways these three agenda items were produced and transmitted in society. This is an essential analysis, as their production and transmission processes later promoted drastic changes in South Korean political geographies around the June 2010 election.

Local vs. Global

School lunch movement debates since 2003 recalled the early 1990s when practical alternatives were promoted to counter the Agricultural Import Liberalization and the Uruguay Round Negotiations of 1993, which legislated the removal of every customs tariff and possible obstacles for trading agricultural products among contracted nations (Kim, 2005). Local Korean economies took a hit as GATT regulations were implemented. Small farmers were forced out of their livelihoods as the global food market expanded, "marching" aggressively into local communities and attracting consumers with cheap, lower quality products. Social activists working closely with local communities asked the school districts to improve the quality of school lunches by using local groceries, which triggered nationwide community development movements inside and outside schools. Using the phrase "locally grown foodstuff" for school lunches challenged GATT regulations of 1994, which strictly prohibit governments from favoring local food producers.

Teachers worked closely with community leadership, social workers, local business entrepreneurs, and parents, asking for their input to remedy the side effects of organization, delivery, and quality that surfaced during the rapid school lunch expansion process (Lee, 2004; K. Choi, 2006). These social movements aimed at creating the momentum to build a local community thriving on collective actions, with schools a central hub. Several goals were woven into school lunch movements: achieving local autonomy and decentralization through grassroots movements; triggering socio-cultural and economic exchanges between rural and urban areas; and developing environmentally friendly, organic agricultural systems connected with local communities (K. Choi, 2006).

In this context, in 2003, the Chonbuk Provincial Assembly enacted ordinances pushing school lunch providers to give priority to local agricultural products in accordance with social workers and local communities' demands. However, the Superintendent of Education in Chonbuk Province appealed to the Supreme Court in January 2004, concerned that the ordinances would violate GATT and later cause future commercial—and diplomatic—conflicts. The Supreme Court decision clarified that provincial offices of education should not give preference to local foods, and that the schools should not favor local groceries to fund lunch (M. Kim, March 10, 2004).

Similar court cases brought by Gyungnam, Gyunggi, Chungbuk, and the Minister of Government Organization and Home Affairs from 2004 to 2005 substantiated the Supreme Court decision in the Chonbuk case (Kim, 2005; K. Choi, 2006). Therefore, social movement organizations and social workers in local communities suggested that "quality agricultural products," "environmentally friendly farm products," or "organic food" would be better words to use in school lunch ordinances so that the schools could have the flexibility to use local farm products while avoiding possible violations of the GATT regulations (Heo & Kim, 2005). Ten provincial assemblies applied such wordings to school lunch ordinances, replacing the term "locally grown farm products" and enabling local groups to keep the spirit of the ordinances supporting local economy. Although a partial victory, the debates around local food sourcing for school lunch were resolved. Later, the local government developed a quality check-up system for local food products (e.g., a G mark for qualified products in Gyunggi Province) so that local food providers could be competitive with global providers and schools would have more choices of locally grown, high quality food products (Goo, 2009).

Opposition to GATT regulations began to gain public consensus (Song, October 10, 2005). Schools' use of local agricultural products became a great model of local self-governing, improving the quality of school lunches and supporting local economies (Kim, 2005). If the school could give a subsidy to buy locally grown products, the community would become self-reliant through the local economy (Kim, 2005). The public began to understand that the competition highly encouraged by GATT provided better conditions for the flow of capital around the globe, not for the economic and political development of all nations (Goo, 2000). South Koreans questioned how to promote public interest when that interest is at the local governments' discretion, often based on the justification that South Korea is a nation under the process of globalization (Kim, 2005). As Kim (2005) mentions, the dilemma is that the stronger a nation's commitment to globalization, the greater the nation's democratic deficits. The importance of global governance based on active local participation of civil society came to the fore.

School-Managed vs. Private-Entrusted Lunch

Since 2002, in addition to concerns about local sourcing of school food, South Koreans have questioned the safety of privatized school lunch systems. Most food poisoning cases were in privatized lunch systems (Moon, November 11, 2002). The quality of school food was poor partly because of the rapid expansion of school lunch programs without adequate governmental funding. The government-funded school lunch facilities were built in stages beginning in 1998 to equip all elementary schools with full lunch facilities. However, executing the middle and high school systems has been more complicated, largely because the government permitted private consignors to serve public school lunch.

Unlike school lunch programs directly managed by schools, serious problems arose as private entrepreneurs pursued profits through the new, guaranteed markets. Private enterprises (e.g., Samsung Everland, LG Ourhome, CJ Food System, Sinsaegae Food System) entered the school lunch market as dominant consignors to profit from gigantic, stable food "markets." The public trusted these leading Korean conglomerates to better serve quality foods. But quality and safety came into question when children in schools with private-entrusted school lunch programs were stricken with severe food poisoning in 2006 (Lee, 2004; K. Choi, 2006).

In the Seoul area alone, 1,557 students in 13 schools suffered food poisoning in 2003 and 3,616 students in 46 schools again suffered food poisoning in 2006. All of these cases occurred in schools with private enterprise lunches (K. Choi, 2006). Considering that only 18.5% of middle school and 34.9% of high school lunch programs are managed by private industry (Korean Industrial Information Center, 2008), the frequent, large-scale outbreak of food poisoning was especially damning of private enterprises' ability to manage school food systems. According to the twelfth clause of Article 2 of the *Food Sanitation Act* (MHW, 2010), school lunch providers should not pursue profits through group food services. However, some consignors spent as little as US$80,000–$150,000 on school lunch facilities, hardly enough to install adequate cooking equipment for commercial kitchens. The consignors sought to maximize their profits to restore their initial investments, so they used inferior quality foods and also bribed teachers and school administrators to renew contracts and to avoid health inspections (Park, September 8, 2003). A large-scale inspection by the Seoul School Health Promotion Center in 1999 reported that 20% of high schools with a private-entrusted school lunch program provided the very lowest level of health safety ("Unsatisfied," 1999). As reports surfaced of private consignors using unidentified food additives and cheap, inferior quality, imported food, parents, educational organizations, and progressive civic groups started a nationwide campaign to improve lunch systems ("Problems," 2002).

The government distributed *The Guidelines of Hygiene Control for School Lunch* (GHCSL) (MHW, 2000a) starting in 2000 and introduced the *Hazard Analysis and Critical Control Point (HACCP)* method (MHW, 2000b) to provide safe, good quality lunches (Cha, 2008). However, as serious safety issues were repeatedly reported, the public demanded that the government change all private-entrusted programs into school-managed ones (Moon, November 11, 2002; Lee, 2004) and persistently challenged school administrators and superintendents to employ school-managed lunch programs that use organic, environmentally friendly foods (Moon, November 11, 2002).

School administrators preferred private consignors because they themselves had very limited funding for school lunch programs and because the private consignors were responsible for the increasingly common unexpected accidents. But the power of social movements led by various civic groups in solidarity could not be ignored. Bureaucratic custom hindered social workers from bringing changes to school lunch programs, but many civic groups challenged those politics, and finally, the National Assembly revised the *School Lunch Law* in June 2006, mandating school-managed lunch at all public and private K–12 schools by January 19, 2010. In accordance with these changes, the Ministry of Education, Science, and Technology published the Comprehensive Countermeasure for Improving School Lunch on December 21, 2006. It guaranteed 97.3% of school-managing school lunch programs by 2009, fully providing students in poverty with free school lunch and renovating dilapidated school lunch facilities with a governmental budget of two billion dollars (K. Choi, 2006).

Free for All vs. Free for Selected Students

Controversies about providing all students with free meals instead of partial funding for targeted students developed into a social consensus as food scares overshadowed social class issues. Free for all school lunch seemed an elusive goal as of late 2009. Yet the number of students who could not pay for school lunches had increased since 2004 (Oh, April 20, 2006). Notably, the number of students who delayed payment or stopped paying for lunch altogether increased by about 88% between 2006 and 2009 even though government-subsidized lunch support for students in need increased by about 38% in the same period, from 526,508 students in 2006 to 730,286 in 2009 (Jang, September 17, 2009). This was because of the polarization of wealth, which has become more and more serious since the foreign currency crisis of 1997. The public also recognized the effects of the new conservative government's recent tax cut policy for the rich (a tax cut of 100 billion dollars) and its massive expenditure for construction around four major rivers (a 22 billion dollar investment) that took away from social welfare programs.

There is a bureaucratic issue regarding the selection process for free lunch recipients, since homeroom teachers are responsible for the recipients' documentation. Students may feel ashamed to submit the required documents (e.g., a certificate of unemployment) to their homeroom teachers, fearing they may experience a "stigma effect" (Cho, 2010)—which has happened when some homeroom teachers mistakenly released free lunch recipients' lists to their students or when a loud "buzzer" was sounded in the cafeteria to reprimand individual students who delayed lunch payment (I. Kim, 2010). Given these instances of ostracism, it became important to find ways to protect students' rights to health and education.

The most radical solution to the problem of public shaming was the "Free Lunch for All" agenda that emerged during the 2006 local government elections. It was a centerpiece of the campaign pledges of the Democratic Laborers' Party, the most progressive radical party in South Korea, which took leadership in 2005 of the revision of the School Lunch Law in cooperation with the School Lunch Network Nationwide and related civic organizations (Jeon, 2006). Their catchphrase "Free Lunch for All" had been silenced until Sanggon Kim, in April 2009, was elected as the Superintendent of Education in Gyunggi Province, the largest province in South Korea and one encompassing metropolitan Seoul. Kim was a joint candidate of a progressive camp in a coalition of about 200 progressive and radical organizations and political parties. As the superintendent-by-election, preparing for a term of office that would last only 14 months, his success would be critical for the restoration of progressive-radical camps. Sanggon Kim promoted "free for all, organic school lunch systems" with the support of the Korean Teachers and Educational Workers' Union and other progressive civic groups, gaining strong public support that later resulted in changes in South Korean political geographies. Kim led radical education and social movements, and his success as a superintendent challenged the hegemonic practices of conservative groups, putting them out of action at the June 2010 election (B. Kim, March 7, 2010).

When Sanggon Kim was elected as Superintendent of Education in Gyunggi Province in 2009, he was surrounded by conservative agents and agencies and had very limited opportunity for practicing progressive educational policies. Kim also encountered serious budget cuts prohibiting the introduction of the "Free School Lunch for All Elementary Students" policy by the Gyunggi Board of Education, whose members mostly supported the ruling Grand National Party and its politics ("Gyunggi," 2009). To make matters worse, the governor of Gyunggi Province and Gyunggi Assembly members all rejected the revised budget plans for providing free lunch to all elementary students (Kim, December 2, 2009). The reasons for conservative groups' rejection of free lunch for all elementary students were clear. First, they argued that funding all students, including economically affluent students, is not reasonable. Second, they contended that education budgets would become bankrupt under "free for all" policies based on "clumsy" populism and socialism (Kim, December 2, 2009).

Conservative decisions provoked bitter public reactions. The opinion section of the Gyunggi Department of Education homepage was filled with objections, blaming the Gyunggi Board of Education members for lack of support. The public objected because (a) providing students with free lunch could be part of compulsory education, which should guarantee all children free tuition, school lunches, school materials, and more; (b) "bankruptcy" of education is an illegitimate claim (to illustrate, Chunbuk, one of the poorest provinces, has the highest free lunch rates in the nation, demonstrating that free lunch is tightly linked to the governor's and the educational leaders' will rather than educational budgets); and (c) feeding students well is not an issue of ideology but of students' human rights to education (Kim, August 11, 2009; Lee, October 16, 2009). In public opinion polls, Kim's free lunch policy won more than 60% approval ratings, while the Board of Education members lost ground, setting up a promising context for progressives in the June 2010 election (Choi, 2010; Son, June 3, 2010).

Meanwhile, the governor of Gyunggi Province, Munsu Kim, argued that "the school was not a soup kitchen" and that funding for recruiting the best teachers and equipping them with better science facilities should be given priority over free lunch systems for all (Kim, December 2, 2009). The conservative educational assembly members echoed him, saying that "free lunch for all" would waste government budgets by benefitting affluent students (Kim, December 2, 2009). In the end, the April 2009 election pledges of Sanggon Kim were invalidated as the majority of educational assembly members, mostly conservatives, rejected budgeting for his free for all school lunch policies. Nevertheless, disputes over the meanings and scope of compulsory education and of social welfare sparked debate, expanding the reach of the arguments. In the next section, I examine how organic, free lunch programs for all students were realized in many provinces through a sophisticated rhetorical battle between conservative and progressive actors.

Free for All, Organic School Lunch Ongoing: The June 2010 Elections and Beyond

The free for all, organic school lunch movements escalated during Sanggon Kim's term of office (May 2009–June 2010), shattering conservative power relations predominant since the 2008 presidential election. But no one (neither conservatives nor progressives) expected the public to show such keen interest in organic, free lunch. As the June 2010 election drew closer, there was speculation as to whether this issue would turn the entire society toward more progressive or radical ways. To understand this important context fully, the philosophies, ideologies, and procedural approaches of both the conservatives and the progressives should be closely examined.

Actors discussed the fundamental philosophies regarding free for all, organic school lunch at many meetings and symposiums held by the School Lunch Network Nationwide and other progressive political parties. Heungsik Jo, one of the discussants for the progressive groups, argued that free school lunch for all is not an option but a required obligation of the government based on the spirit of the UN Convention of the Rights of the Child. According to the 6th clause of the UN Convention, the government should guarantee children a maximum of rights to live and grow, he reminded attendees. Jo also mentioned that as the school and government do not run a "dining enterprise," they should not promote environments where economically marginalized students skip meals because they lack money (Jeong, 2010). Jo's ideas corresponded to those of many others supporting the concept of universal social welfare—that is, providing welfare programs to every person in society regardless of her or his socioeconomic status. It is a more aggressive concept than selective social welfare, which gives charity to socio-economically marginalized people only if they prove their status (Choi, 2010). Under universal social welfare systems, beneficiaries of school lunches are all minors, so selective school lunch policies based on their *parents'* income, region, age, and gender should not be accepted (Heungsik Jo, 2010, as cited in Jeong, 2010). Adding to the discussion, Bae (March 2010), the representative of the National Public Movements Camp for Safe School Lunch, made a case for the importance of free school lunch in an era of economic crisis when the layer of the middle class had become thinner and people's livelihoods were diminished. Bae (March 2010) argued that the increasing number of students missing meals and delaying lunch payments should be scrutinized; he urged the government to fund free school lunch for all students so that they would no longer experience educational discrimination because of their economic circumstances.

The scope of obligatory education was discussed at many seminars and in journal articles. For instance, the National Camp for School Lunch Movements held a symposium to promote free for all, organic school lunch in March 2010. The symposium included the participation of elites from various circles, and it initiated signature-collecting campaigns in collaboration with about 500 civic groups nationwide (The Gyunggi Camp for Free for All, Organic School Lunch Movements, 2009). The School Lunch Network Nationwide, Minseok Ahn (a National Assembly member from the Democratic Party), and Sanggon Kim (Gyunggi Superintendent of Education) held a symposium at Hwasung Office of Education in Gyunggi Province for the public to understand the current circumstances surrounding school lunch movements, supported by local broadcasting systems and the Agricultural Cooperatives (The Gyunggi Camp for Free for All, Organic School Lunch Movements, 2009). At this meeting, Ahn (2009) explained the meaning of "compulsory education," citing Articles 2 and 3 of the 31st clause of the Constitution, and argued that an obligatory education should provide all children with free education programs including free school lunch. Goo (2009) also stated that obligatory education should cover all school materials, textbooks, and school lunches as well as tuition, and every student should have a right to free school meals without experiencing discrimination. On the other hand, Lee (2010, March 12) saw this issue from an economist's perspective, arguing that obligatory education is one of the "merit goods" that the government needs to supply for all citizens to receive a minimum level of benefits. According to Lee, supplying free lunch to economically affluent students might not be appropriate from the perspective of free lunch as social welfare, but free lunch for all should be reasonable from the perspective of "free lunch as merit goods" because the government should provide merit goods for free to all. The rich do not pay for their children's school lunches separately, but they would pay more taxes to cover the costs, so the free for all school lunch policy is not against equity and fairness (Lee, 2010, March 12). Even though his interpretation on social welfare is narrow, Lee's perspective served to refresh the people's minds and triggered dynamic debates.

Conservatives, including the ruling party and many civic groups supporting President Myungbak Lee and his government, took an opposite stance on social welfare, obligatory education, and free school lunch policy. Two editorials in the most conservative newspapers in South Korea ("Competition," 2010, February 4; "Voters," 2010, February 4) said that politicians and candidates backing free for all school lunch would distort the allocation of educational resources and split public opinions and that this would lead to a serious national economic crisis ("Competition," 2010, February 4; "Voters," 2010, February 4). Concerned that the pledge of the free for all school lunch had spread beyond Gyunggi Province, these editorials argued that smooth-talking progressive politicians would sprout up like toadstools from under the shadow of inequality and economic disparity. Gusik Choi, a National Assembly member of the ruling Grand National Party, argued that giving free meals to affluent students would be possible in an earthly *paradise*, but that attempts to do so in reality reflected mere populism and would cheat the citizenry (Kim & Lee, March 12, 2010). President Myungbak Lee joined these disputes, saying that free school lunch for all goes against the philosophy of selective social welfare and arguing that the rich would need to pay to help ordinary people (Kim & Lee, March 12, 2010). The Grand National Party again argued that the free lunch policy amounted to an indiscriminate populism, which did not consider reasonable ways of securing finances.

Progressive radical parties and civic groups agreed to a gradual expansion of the free lunch policy, citing concern for national finances, specifically deep tax cuts for the rich and the national debt. Progressives wanted to phase in the free lunch policy expansion by grade levels: grades 1–3 in 2011, 4–6 in 2012, and 7–9 in 2013. Conservatives, on the other hand, wished to phase in the program based on students' living environments, or by regions and family income. Under the conservative plan, more students in socio-geographically isolated areas from families with limited incomes would receive free meals. The government guaranteed an increased rate of free meal beneficiaries from 13% of students at the bottom of the economic ladder (based on the national income index by head of houschold) in 2010 to 26.4% in 2012 (Kim & Lee, March 18, 2010). The government announced that the combined online social welfare network of the Ministry of Health and Welfare would directly notify the schools of the free meal beneficiaries to simplify the application process and prevent private information from being released to the public (Kim & Lee, March 18, 2010). As the disputes surrounding free school lunch for all became public, conservatives became increasingly concerned over the possibility of losing the June 2010 election. In response, the government and the Grand National Party adopted a free day care policy promising full day care expenses (for children 0–5 years old) and preschool costs (for children 3–5 years old) for families in the 70th percentile or lower of income level. Such ideas, however, did not gain much support. The progressive parties merely confirmed the possibility of the free lunch policy by reducing unnecessary governmental spending on public relations (Kim & Lee, March 18, 2010).

Achieving widespread public consent for the expansion of universal social welfare, an alliance of 4 oppositional parties and 2,100 civic groups and grassroots organizations won the June 2010 election in most provinces nationwide including Seoul and Gyunggi Province. The Democratic Party, the first oppositional party supported by relatively progressive groups, largely swept the election, obtaining a total of 60–70% of the seats on Boards of Education and

Provincial Assemblies. Six of sixteen newly elected Superintendents of Education were leaders of human rights movements, including movements for students' rights in schools and school innovation movements. This makeup indicated that the socio-political turn toward conservativism after the 2008 Presidential Election was waning (Son, June 3, 2010). South Koreans have four years to think more seriously about the welfare state. The progressive radical victory has opened promising space for the realization of free for all, organic school lunches in South Korea. Based on election promises, plans for free for all, organic school lunch systems will be designed by the Superintendents of Education. Education budgets will comprise a 50:50 match of funds from the Local Boards of Education and the Provincial Assemblies. Free for all, organic school lunch policy systems are now likely to be promoted in many provinces and cities (see Figure 1), even though Gyungbuk Province and two megalopolises (Ulsan and Daegu) wish to retain more conservative lunch policies, and the policies of Gyungnam Province and Daejeon megalopolis are still being negotiated (Choi, 2010), partly because of ideological differences between these provinces' governors and superintendents of education.

As a majority of Superintendents of Education announced their plans to roll out free for all, organic school lunch in stages from 2011 to 2014, the public has begun to reconsider local, environmentally friendly, organic school meals and school-managed lunch systems that would provide safer and healthier meals than the private-entrusted ones. Still, the local-global treaty conflict remains, particularly in light of the South Korean and U.S. governments' ratification of the Free Trade Act in 2008, which contained strong regulations restricting any governmental boosting of local farming and markets, including each government's subsidies of local economies. The 2008 agreement stipulated that schools should not prefer locally grown groceries and meats. To counter, educational administrators began to develop standards and guidelines to guarantee the quality of food supplies (e.g., G-mark by the governor of Gyunggi Province for Qualified Food Supplies). However, the word "local" has been removed from official documents and social agendas, marking a win for globalized agricultural markets and their politics, armed with international treaties and contracts against small-scale local farming and local markets. Struggles for school-managed lunch has proven a more complicated issue, though the School Lunch Law, revised in 2006, mandated a change from private-enterprise secondary school lunch systems to school-managed ones.

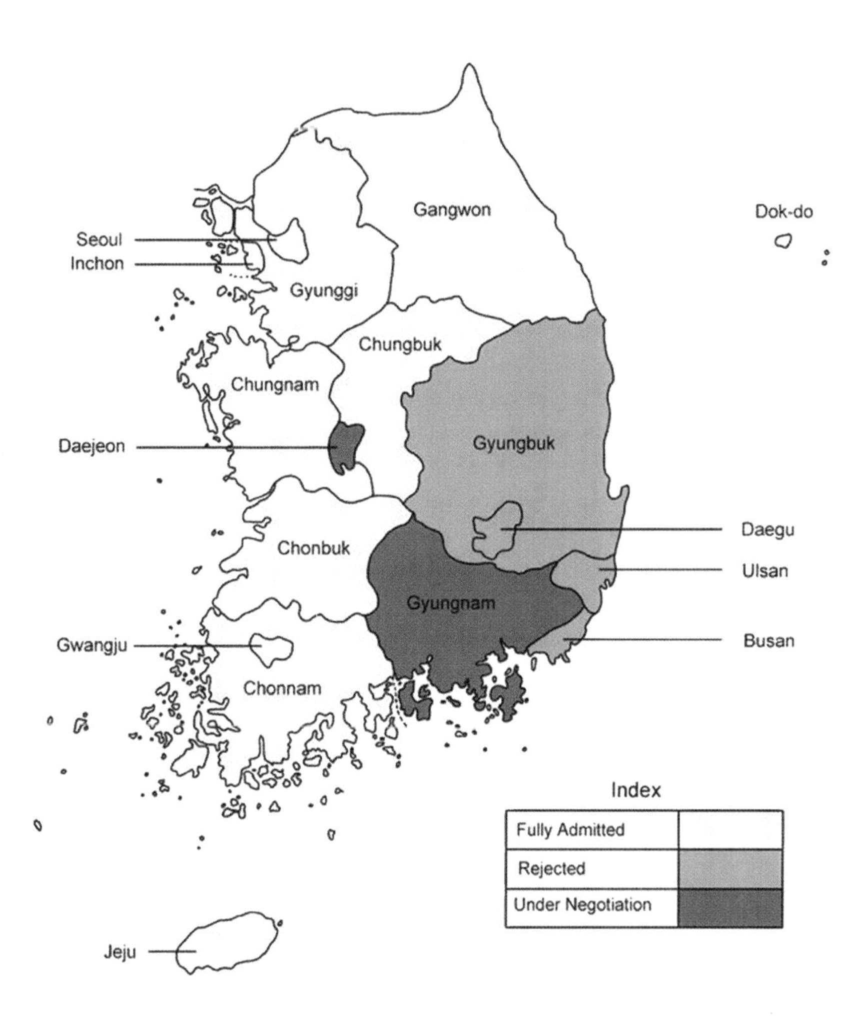

Figure 1. Map of Free for All School Lunch Distribution (as of June 2010)

Note. This map is based on the June 2010 election. Though some political leaders of small cities and counties have reneged on their support of free for all school lunch systems, in general the school lunch policies will be promoted as shown by 2014.

South Korean society opened another door to social welfare policies with the June 2010 election outcomes. In the early 1990s, local community development movements began advocating for school lunch plans using locally grown food, and many agents and agencies with different political perspectives have united in support of a policy to make organic school lunches free for all students. Free for all, organic school lunch systems may be a reality for all of South Korea's students by 2014.

Conclusion

The free for all, organic school lunch system will be introduced in most provinces and cities in late 2010. In this policy context, the public and not the educational administrators and state governors were committed to bringing drastic changes to the school lunch systems as well as to social welfare programs. In fact, 2,100 civic groups and social organizations worked together to support progressive radical parties and voted for the candidates who promised to spend more money on social welfare programs. Controversy remains about how to enhance local economies without violating global trade regulations and contracts. Breaking up privatized systems and moving toward publicly managed ones also is a complicated issue as new policies could violate private consignors' rights. Despite these issues, South Koreans continue to press for socio-ecological change, challenging the government to better serve students and the general public.

The South Korean case of implementing free for all, organic school lunch systems gives policymakers and educational researchers abroad a greater sense of the complexity of the school food policymaking process, which involves diverse layers of stakeholders from politicians to parents to business entrepreneurs to local farmers (Morgan & Sonnino, 2008). Understanding specific school lunch models like this one is helpful for policymakers who want to find ways to work through such conflicts by negotiating across their own socio-cultural, ecological, and political contexts (Ball, 1998).

References

Ahn, J. (1992, March 18). Debates on financing school lunch systems. *Hangyorae Newspaper*, 8.

Ahn, M. (2009). Students' happy lunchtime is our responsibility. In M. Ahn (Chair), *How to successfully implement free for all school lunch system.* Symposium conducted at the meeting hall of Gyunggi Hwaseong Office of Education, South Korea.

Bae, O. (2010, March). Three goals and ten tasks for organic, free for all school lunch. In J. Choi (Chair), *Why organic, free for all school lunch?* Symposium conducted at the 59th Participatory Forum of Peoplepower21, Zelkova Hall, Seoul, Korea.

Baker, B. D., & Richards, C. E. (2004). *The ecology of educational systems: Data, models, and tools for improvisational leading and learning.* Upper Saddle River, NJ: Pearson Education.

Ball, S. J. (1994). *Education reform: A critical and post-structural approach.* Buckingham, England: Open University Press.

Ball, S. J. (1998). Big policies/small world: An introduction to international perspectives in education policy. *Comparative Education, 34*(2), 119–130.

Cha, H. (2008). *A literature review on HACCP school food service* (Unpublished master's thesis). Graduate School of Education, Kyeongsan University, South Korea.

Cho, H. (2010). The appropriateness of organic, free for all school lunch and its implementation strategies. In J. Choi (Chair), *Why organic, free for all school lunch?* Symposium conducted at the 59th Participatory Forum of Peoplepower21, Zelkova Hall, Seoul, Korea.

Choi, C. (2010, July). Realize indiscriminate, universal welfare in education and compulsory education through the free for all school lunch programs! In M. Ahn, Y. Kweon, S. Lee, NMCSSL, and Future Education Hope (Chairs), *A promise with citizens: Action plans for realizing organic, free for all school lunch programs.* Open forum conducted at the meeting hall of the National Assembly, Seoul, Korea.

Choi, H. (2006, December 21). 22 billion U.S. dollars promised for school lunch systems. *Hangyorae Newspaper*, 3.

Choi, K. (2006). *Problems of school lunch and directions of school lunch movements by civic organizations* (Unpublished master's thesis). Graduate School of NGO Policies, Hanil Jangsin University, Seoul, Korea.

Choi, Y. (2010, June 3). Seoul, Possible to implement free for all school lunch from next year. *Maeil Economy*. Retrieved from http://news.mk.co.kr/v3/view.php?year=2010&no=286770

Competition for the free for all school lunch is a toadstool before the 2010 election [Editorial]. (2010, February 4). *Chosun Daily*. Retrieved from http://news.chosun.com/site/data/html_dir/2010/02/03/2010020301930.html

Goo, C. (2000). *Globalization, reality or another myth?* Seoul, Korea: Chaeksaesang.

Goo, H. (2009). Tasks for realizing organic, free for all school lunch. In M. Ahn (Chair), *How to successfully implement free for all school lunch system.* Symposium conducted at the meeting hall of Gyunggi Hwaseong Office of Education, South Korea.

Goodlad, J. I. (Ed.). (1987). *The ecology of school renewal.* Chicago: National Society for the Study of Education.

Gyunggi Camp for Free for All, Organic School Lunch Movements. (2009). *How to make success of free for all, organic school lunch?* Symposium booklet. Gyunggi, South Korea.

Gyunggi Local Board of Education is taking away children's rights for meals [Editorial.] (2009, June 25). *Hangyorae Newspaper*, 23.

Hall, P. (1986). *Governing the economy.* Cambridge, UK: Polity Press.

Heo, M., & Kim, T. (2005, September 10). It is inevitable to drive school lunch policies towards the ones that the schools directly manage. *Hangyorae Newspaper*, 8.

Jang, S. (2009, September 17). The numbers of students who did not pay for their lunch money rapidly increased over the years. *Nae-il Newspaper*, 19.

Jeon, B. (2006, May 2). Choice, 05/31/09: The hotly contested era of regional elections. *Gyunghyang Newspaper*, 6.

Jeong, H. (2010, February 20). An interview with Heungsik Cho: Free for all school lunch is the government's obligation and fits the UN Convention for children's education. *Gyunghyang Newspaper.* Retrieved from http://news.khan.co.kr/kh_news/khan_art_view.html?artid=201002200200455&code=910110

Ji, W. (1994, November 8). School lunch budgets for the next school year were omitted. *Saegae Daily*, 11.

Kim, B. (2010, March 7). Sanggon Kim Fobia. *Gyunghyang Newspaper.* Retrieved from http://news.khan.co.kr/kh_news/khan_art_view.html?artid=201003071835165&code=990507

Kim, I. (2010). Students dash to the cafeteria at the speed of light when the lunch bell rings. In J. Choi (Chair), *Why organic, free for all school lunch?* Symposium conducted at the 59th Participatory Forum of Peoplepower21, Zelkova Hall, Seoul, Korea.

Kim, K. (2009, December 2). Munsu Kim, free for all school lunch represents populism. *Yeonhap News.* Retrieved from http://news.naver.com/main/read.nhn?mode=LSD&mid=sec&sid1=100&oid=001&aid=0003005367

Kim, K., & Lee, I. (2010, March 12). The ruling party, "school lunch for *chaebols* [extremely rich groups]" vs. the oppositional party, "social welfare with stigma/labeling." *Gyunghyang Newspaper.* Retrieved from http://news.khan.co.kr/kh_news/khan_art_view.html?artid=201003120223305&code=910402

Kim, K., & Lee, I. (2010, March 18). The government and ruling party urgently offered selective school lunch policies against the oppositional parties' free for all policies. *Gyunghyang Newspaper.* Retrieved from http://news.khan.co.kr/kh_news/khan_art_view.html?artid=201003181819545&code=910110

Kim, M. (2004, March 10). Current condition of the enactment for the school lunch ordinances by Chounbuk Department of Education. Retrieved from http://blog.peoplepower21.org/Welfare/12699

Kim, T. (2005). Globalization of economic administration. *Law,* 46(4), 312–348.

Kim, Y. (2009, August 11). Free for all school lunch is not an issue of ideology. *Gyunghyang Newspaper*, 56.

Korean Industrial Information Center. (2008). *Korean educational statistics annual.* Seoul, Korea: Publication Committee for the Korean Educational Statistics Annual.

Lee, B. (2004). *The directions for school lunch movements and suggestions for local practices.* Unpublished manuscript, Seoul, Korea.

Lee, I., & Hong, S. (1998, April 4). Students without lunch packs increased. *Dong-A Daily*, p. 23.

Lee, J. (2005). Korean education fever and private tutoring. *KEDI Journal of Educational Policy,* 2(1), 99–107.

Lee, J. (2010, March 12). *A comment regarding current debates on school lunch programs.* Retrieved from http://jkl123.com/sub3_1.htm?table=my1&st=view&page=1&id=91&limit=&keykind =&keyword=&bo_class=

Lee, S. (2009, October 16). Is the expansion of free lunch a socialistic idea? *Oh My News*. Retrieved from http://www.ohmynews.com/NWS_Web/view/at_pg.aspx?CNTN_CD=A0001316459

Levine, S. (2008). *School lunch politics: The surprising history of America's favorite welfare program*. Princeton, NJ: Princeton University Press.

Ministry of Education and Human Resource (MEHR). (1997). *The school lunch law*. Seoul, Korea.

Ministry of Education and Human Resource (MEHR). (2006). *The school lunch law*. Seoul, Korea.

Ministry of Education. (1981). *The school lunch law*. Seoul, Korea.

Ministry of Education. (1993). *The school lunch law*. Seoul, Korea.

Ministry of Health and Welfare (MHW). (2000a). *The guidelines of hygiene control for school lunch*. Seoul, Korea.

Ministry of Health and Welfare (MHW). (2000b). *Establishing the hazard analysis and critical control point (HACCP) system for the improvement of school food service safety*. Seoul, Korea.

Ministry of Health and Welfare (MHW). (2010). *The food sanitation act*. Seoul, Korea.

Moon, S. (2002, November 11). School lunch renovation movements ignited to revise the School Lunch Law and to vindicate its honor after a hotbed of food poisoning. *Hangyorae Newspaper*, 34.

Morgan, K., & Sonnino, R. (2008). *The school food revolution*. London, England and Sterling, VA: Earthscan.

National Movement Camp for Safe School Lunch (NMCSSL). (2003). www.geubsik.org

Oh, C. (2006, April 20). 5,000 more students did not pay for their school lunch than last year. *Gyunghyang Newspaper*, 10.

Park, B. (2003, September 8). *Human rights of students and school lunch*. Retrieved from http://blog.peoplepower21.org/welfare/9291

People's Solidarity for Participatory Democracy (PSPD). (1994). Retrieved from www.peoplepower21.org

Poppendieck, J. (2010). *Free for all: Fixing school food in America*. Berkeley: University of California Press.

Problems of school lunch programs by private-entrusted systems should be removed [Editorial]. (2002, November 5). *Hangyorae Newspaper*. Retrieved from http://www.hani.co.kr/section-001001000/2002/11/ 001001000200211051900144.html

School Lunch Network Nationwide (SLNN). (2002). Retrieved from www.schoolbob.org

Sim, K. (1992, September 15). Burdens of fully financing elementary school lunch programs nationwide. *Dong-A Daily*, 9.

Son, B. (2010, June 3). The oppositional party fully won the Provincial Assembly elections around Seoul area. *Oh My News*. Retrieved from http://m.ohmynews.com/NWS_Web/Mobile/at_pg.aspx?cntn_cd=A0001394111

Song, H. (2005, October 10). Most economically developed countries provide with school lunch with locally grown foods. *Gyunghyang Newspaper*. Retrieved from http://news.khan.co.kr/kh_news/khan_art_view.html?artid=200510101603241&code=900305

Unsatisfied health administration of the school lunch system [Editorial.] (1999, August 11). *Gukmin Daily*, 6.

Voters should accept reality of "free for all school lunch" as it is [Editorial]. (2010, February 4). *Dong-A Daily*. Retrieved from http://news.donga.com/3/all/20100204/25927918/1

Weaver-Hightower, M. B. (2008). An ecology metaphor for educational policy analysis: A call to complexity. *Educational Researcher, 37*(3), 153–167.

World Trade Organization (WTO). (1994). *General Agreement on Tariffs and Trade.* Retrieved from http://www.wto.org/english/docs_e/legal_e/06-gatt_e.htm

Yoo, I. (1995, April 23). Urge the government to pay for all feeding facilities of school lunch systems. *Gyunghyang Newspaper*, 12.

• SECTION TWO •

Reforming School Food

Parents, Activists, Teachers, and Youth

• CHAPTER SIX •

School Food, Public Policy, and Strategies for Change*

Marion Nestle

School food is a "hot button" issue, and it well deserves to be. It lies right at the heart of issues related to equality in our society. Americans live in a pluralistic society. For democracy to work, the interests of constituencies must be appropriately balanced. School food is about the balance between corporate interests and those of advocates for children's health.

The nutritional health of American children has changed during this century, improving dramatically in some ways, but not in others. In the early 1900s, the principal health problems among children were infectious diseases made worse by diets limited in calories and nutrients. As the economy improved, and as more was learned about nutritional needs, manufacturers fortified foods with key nutrients, the government started school feeding programs, and the results were a decline in nutrient deficiency conditions. That severe undernutrition has now virtually disappeared among American children can be counted as one of the great public health achievements of the twentieth century. For the great majority of American children, the problem of not having enough food has been solved. Whether children are eating the right food is another matter.

Indeed, the most important nutritional problem among children today is obesity—a consequence of eating too much food, rather than too little. Obesity rates are rising rapidly among children and adolescents, especially those who are African-American or Hispanic. The health consequences also are rising:

---

* "School Food, Public Policy, and Strategies for Change" by Marion Nestle was originally published by the Center for Ecoliteracy. For more information, visit www.ecoliteracy.org.

high levels of serum cholesterol, blood pressure, and "adult-onset" diabetes. This increase has occurred in response to complex societal, economic, demographic, and environmental changes that have reduced physical activity and promoted greater intake of foods high in calories but not necessarily high in nutrients.

This shift—from too little to too much food—has created a dilemma for the United States Department of Agriculture (USDA), other federal agencies, and many of my fellow nutritionists. Since its inception, the USDA has had two missions: to promote American agricultural products and to advise the public about how best to use those products. The school lunch program derived precisely from the congruence of the two missions. The government could use up surplus food commodities by passing them along to low-income children. As long as dietary advice was to eat more, the advice caused no conflict.

Once the problems shifted to chronic diseases, however, the congruence ended. Eat less means eating less of fat, saturated fat, cholesterol, sugar, and salt, which in turn means eating less of the principal food sources of those nutrients—meat, dairy, fried foods, soft drinks, and potato chips. USDA was then faced with the problem of continuing to promote use of such foods while asking the public to eat less of them—a dilemma that continues to the present day.

For the federal government to suggest that anyone eat less of any food does not play well in our political environment; such suggestions hurt sales. This matters, because we vastly overproduce food in this country—a secret seemingly known only to analysts in the Economic Research Service. The average per capita supply of calories available from food produced in the United States—plus imports, less exports—is 3,900 per day for every man, woman, and child, more than twice what is needed on average. These are food availability figures and they cover food wasted, fed to pets, and fats used for frying, but they have gone up by 600 calories since 1970 and are more than sufficient to account for rising rates of obesity among adults and children. The point here is that overproduction makes for a highly competitive food supply. People can only eat so much. So to sell more, companies must get us to eat their foods rather than those of competitors, or to eat more, thereby encouraging us to become obese.

The stakes are very high. Food is a $1.3 trillion annual business, with the vast percentage of profits going to added-value products rather than basic commodities. It pays to turn wheat into sugared breakfast cereals, or potatoes into chips. Farmers get only a small share—18% or so—of the consumer's food dollar, less for vegetables, fruits, and grains than for meat and dairy. So there is a big incentive to marketers to make food products with cheap raw ingredients like fat and sugar. And they do—to the tune of 12,000 or so new products every year. There are now 320,000 foods on the market (not all in the same place); the average large supermarket contains 40,000 to 60,000 food products, more than anyone could possibly need or want.

This level of overproduction has kept growth in the food industry stagnating for years at about a 1% growth rate, far lower than in comparable industries. Corporations need better growth rates than that to satisfy shareholders. To expand sales, food companies can try to sell products overseas, or they can try to increase market share at home. With this understood, it is evident why marketers so relentlessly pursue children as potential sales targets: children 7 to 12 years of age spend billions of their own money on snacks and beverages, and teenagers have billions more to spend on candy, soft drinks, ice cream, and fast food—precisely the types of foods that promote high caloric intakes. The influence of children on adult spending is even greater. Kids are said to influence one-third of total sales of candy and gum, and 20% to 30% of cold cereals, pizza, salty snacks, and soft drinks.

Food companies say they are not responsible for the changes in society that make kids demand their products. They point to decreasing family size, older parents, working parents, and single parents all predisposing to greater indulgence of kids. Kids are more spoiled; coupled with other changes in society, they are also less independent. From the age of eight on, my friends and I could and did take New York subways by ourselves, a level of independence utterly inconceivable today. Parents want their kids to make independent choices whenever they can, and foods are perfect opportunities for such decision-making, which is just what marketers want.

Of course, what kids are doing instead of taking subways is watching food commercials on TV or on the Internet. Advertisers quite unapologetically direct marketing efforts to kids as young as six. They consider this targeting quite sensible. And they know how to do it. The research available to help advertisers target children is awe-inspiring in its comprehensiveness, level of detail, and thoroughly undisguised cynicism. Not only have marketers identified precisely the kinds of packages and messages most likely to attract boys, girls, or kids of varying ages, but they also justify advertising to children as a public service.

The USDA is a complex agency with multiple constituencies. Because of internal conflicts of interest, the agency cannot protect the integrity of the school meals program on its own. It is already clear, for example, that Congress believes that more competition is good for schools. If USDA wants to help children prevent obesity through healthier school lunches, it needs to be working with a much broader set of allies. USDA cannot tell children to eat less of any food, and the school meal programs still reflect the dual goals of their origin. My recommendation would be to enter into partnership with the Department of Health and Human Services and the Department of Education to develop an interagency alliance for a national school health campaign focused on obesity prevention using the Healthy People 2010 goals as a starting point.

What particularly disturbs me about commercial intrusions into school meals is that they are so unnecessary. Schools are perfectly capable of producing nutritionally sound foods that taste good and are enthusiastically consumed by students as well as teachers. From my own observations, a healthy school meals program (in every sense of the word) requires just three elements: a committed food service director, a supportive principal, and devoted parents. It just seems so obvious that the future of our nation demands each of these elements to be in place in every one of the 95,000 schools in the country. These are, after all, our children.

There needs to be one place in society where children feel that their needs come first—not their future as consumers. In American society today, schools are the only option. That's why every aspect of school food matters so much and is worth every minute spent to promote and protect its integrity.

•CHAPTER SEVEN•

Food Prep 101
Low-Income Teens of Color Cooking Food and Analyzing Media

Catherine Lalonde

As cheap, fast food restaurants proliferate and the low nutritional content of fast food moves into supermarkets via prepackaged meals (i.e., White Castle burgers in the frozen foods section), food preparation skills are on the downslide among growing portions of consumers worldwide. Further, the endless drive for profit pushes companies to target younger populations of children, now wooing even 2-year-olds through glossy, high-movement and colorful ad campaigns (Nestle, 2007). As schools begin to address rising obesity and adult onset diabetes among their student populations (and among low-income students, specifically), already-strapped budgets are overtaxed by these efforts to increase both physical activities and nutritional education. Nutritional researchers and urban activists are tracking foods consumed in or accessible through after-school programs (Epstein et al., 2006; Granner et al., 2004; Raja, Born & Kozlowski, 2008; Suarez-Balcazar et al., 2006; see also Sandler, this volume), while others are conducting nutrition education interventions (Mozaffarian et al., 2010; Rinderknecht & Smith, 2004) and focus groups to learn about food choice influences (Baer Wilson, Musham & McLellan, 2004; Cunningham-Sabo et al., 2008; Fontenot Molaison et al., 2005). What might educators and activists do outside of schools to address food preparation and advertising issues, and what might such an intervention look like? Further, how might afterschool interventions influence a healthier school food "policy ecology" (Weaver-Hightower, 2008)?

In this chapter, I describe and analyze my experiences while working with an after-school program located in the northeastern United States with 13- to 17-year-old African American and Hispanic low-income students as they learned to prepare food during a cooking course and to question the media messages about the food they consume. As this course was conducted only once per week for two hours over five weeks, it represents a pilot study for further research about how to blend food preparation and analysis of food-related media messages into an after-school program. I designed this course in relation to a critical pedagogy framework (Freire, 1970, 2005; hooks, 1994; Kincheloe, 2005), integrating skill sets and teaching approaches inspired by "slow food" activists, nutritionists, and food theorists (Albon & Mukherji, 2008; Berry, Davidson & Grey, 2004; Bisset & Potvin, 2007; Izumi et al., 2006; Kolbe, Kann & Brener, 2001; Morgan & Sonnino, 2008; Murray, 2007; Nestle, 2007; Petrini, 2007; Pollan, 2008; Roberts, 2008; Rosecrans et al., 2008).

I begin with a brief look at pertinent literature, followed by a contextual overview of the community center and cooking course. Next, I explore the three challenges that emerged during the pilot project: disrupting the gendered dimensions of cooking, understanding food relationships through the lenses of religion and food sensitivity issues, and balancing the highly structured environment of the community center with the open, critical pedagogical approach of the cooking course. Finally, I consider the implications of this course and the challenges that emerged, and I speculate about what future cooking courses and other research might explore. Throughout this analysis, I reflect on my experience as a critical educator and on how this course functioned within the context of a community center as a guiding example for similar future interventions.

Food Preparation as a Critical Pedagogy Intervention

Critical pedagogy involves continuous reflection and action in relation to one's surroundings (Freire, 2005)—and in this case, in terms of one's use and awareness of food knowledge and preparation. In critical pedagogy, students must be empowered by problem-posing rather than "banking" concepts of education (Freire, 1970), constantly questioning their relationships with what they consume, what they've come to recognize as "food," and what processes their food undergoes before being consumed. Further, critical educators mobilize a sense of urgency in educating for social justice, encouraging their students to turn these critical lenses on their local and global communities in an effort to resolve structural issues (Kincheloe, 2005). By raising awareness about food production, preparation and consumption issues, students are not only in tune with their local community's food-related practices, but they also foster a consistent awareness of how these local practices impact and are impacted by global processes.

As a critical researcher, it is important to situate myself in this process. I am a white, middle-class woman with no formal training in food preparation but many years of experience cooking for family, friends, and myself. Trained in the social foundations of education and media studies, I have researched how social issues impact identity formation (Weis et al., 2010) and how media images impact these same critical constructive processes (Lalonde, in press). I engaged in a critical examination of food preparation and food-related media images in an after-school intervention in a community center as a critical educator who urgently wished to act.

Issues raised and explored by various food activists, theorists, and nutritionists can stimulate critical pedagogical action with students of color from low socioeconomic backgrounds. One major issue is the marketing of food products to children. Marketing towards children moves beyond television and "counting games" for schools devised and distributed by snack companies; food corporations have integrated their products into school vending machines and lunch programs, garnering the greatest amount of interest from underfunded schools attended predominantly by students of color from low socioeconomic backgrounds (Nestle, 2007). This intense marketing of ever more processed, packaged, and mobile food has resulted in decreased nutritional knowledge and interest in developing food preparation skills in the home, all in favor of the "fast, easy, and cheap" food products whose greater fat and sugar contents (and far lower nutritional content than found in "whole" foods like fruits and vegetables) have led to rising health care costs to treat diseases resulting from this "Western diet" (Murray, 2007; Pollan, 2008; Roberts, 2008). Some activists are resisting these trends through organizations such as the Slow Food Movement (Petrini, 2007). This food-related literature suggests the importance of generating food preparation skills as a way of addressing the socioeconomic pressures resulting from "going out to eat" or otherwise depending on corporate food enterprises that are not only expensive but also possibly detrimental to one's health. Using this perspective, an intervention with teens at a community center would need to include both food preparation skill development and critical media analysis of food-related images to help disrupt status quo ideas of what constitutes "food" and how to prepare it.

Much research investigating links between education and food issues focuses on either addressing health issues like diabetes (Berry, Davidson & Grey, 2004; Rosecrans et al., 2008) or are mixed-method studies (Bisset & Potvin, 2007; Raja, Born & Kozlowski, 2008) or syntheses of regional, statewide or national quantitative studies of school-based nutritional programs (Epstein et al., 2006; Izumi et al., 2006; Kolbe, Kann & Brener, 2001). In this chapter, I use qualitative methods (described below) to examine an intervention involving food preparation and discussion of food-related media images with teens in a community center. Qualitative methods not only help us better understand how teens approach food preparation and how they process images of food, but also help us gain insight into how we might disrupt unhealthy eating and facilitate connections with healthy dietary practices. Furthermore, "at risk" students (Berry, Davidson & Grey, 2004), First Nations students (Rosecrans et al., 2008), and public school students in regions of the United States and United Kingdom (Albon & Mukherji, 2008; Morgan & Sonnino, 2008) are discussed in current literature or part of focus group research (Baer Wilson, Musham & McLellan, 2004; Burnet et al., 2007), this pilot focuses on low socioeconomic children and teens of African American or Hispanic backgrounds.

In the context of a school food policy ecology, this cooking class intervention constitutes a more local and individual layer, which is no less important than the federal and international levels, and involves its own "texts, histories, people, places, groups, traditions, economic and political conditions" (Weaver-Hightower, 2008, p. 155). It is important to bear in mind the possible impacts that critical pedagogical interventions, though highly individual, can have on the lives of the teens—and instructors—involved in them.

Cooking Class: Researching and Working in the Kitchen

Prospect Clubhouse—this and all names are pseudonyms—is located on a side street near one of the city's highways and next door to a church, and it is far enough into the block to be removed from the highway's noise. In between the church and the actual clubhouse is a fenced-in playground and picnic area where students frequently eat their meals when it is not raining. Upon entering Prospect Clubhouse, I am greeted with the laughing and screaming of young children (the "8–13-year-old group" I am later informed) for whom staff members are preparing lunches. Some children are being relocated to the back of the lunch line; I hear that ham sandwiches are lunch for everyone else but not for them, due to what might be religious and/or dietary restrictions. I file this away to investigate later. A search for local census information yields population numbers relating to race, gender, income levels, and spoken language, yet does not provide hard data about religious affiliations in the area.

A combination of pictures and decorations made by the children (including a long paper dragon hung from the ceiling), educational posters mobilizing mainly African American leadership quotes and portraits, and posters of past recipients of "club member of the year" awards cover the walls of the main clubhouse room. Mr. N., as both the staff and youth call the community center director, is seated behind a computer (that looks to be from the late 1980s). He shakes my hand and welcomes me after I walk in. He notes that the kids are about to have lunch, after which I will be able to set up in the kitchen and begin class.

This introduction to my work with the teens and staff at Prospect Clubhouse provides a window into the qualitative methods used. I combined action research approaches (Fine et al., 2004; Foreman-Peck & Murray, 2008) with focus grouping techniques (Johnson, 1982; Morgan, 1997) to properly understand the meanings made by the African American and Hispanic teens from low socioeconomic backgrounds that participated in the after-school cooking program. I used Participatory Action Research (PAR) methods with a focus on "critical social science" approaches (Foreman-Peck & Murray, 2008) to involve the teen participants in constructing the recipe lists and resulting meals that we prepared together, as well as consistently posing questions to teens about food-related media throughout the intervention. The critical elements of "participation *with*, not only *for*, community" (Fine et al., 2004, p. 98, emphasis original) and "theoriz[ing] and strategiz[ing] how PAR gives back to communities good enough to open themselves up for intellectual scrutiny" (Fine et al., 2004, p. 99) guided this research and the reflection undertaken in this chapter.

Although formal focus group opportunities were limited during this intervention, Johnson's (1982) work helped support my efforts to interact with small groups of teens—three or four students—while preparing food, which offered a more informal and intimate interaction with the teens while we worked together on meals. Since the goal of this work and this chapter is to reflect on my experiences throughout this food preparation and food-related media analysis intervention, most of my data is derived from my observations of the teens during this process.

As noted above, this work is of a pilot-study scale, and, as such, the format and techniques are continuously considered in order to address project unknowns for future research. The broader research project will involve similar food preparation and media analysis courses at multiple community center sites, and the project goal is not only to produce increased cooking and media analysis skills among the teens involved, but also to begin building connections with schools interested in creating similar programs for their students. Ultimately, it is my hope that the flexible, innovative methods associated with the critical pedagogical approach to this course will be used with students in pre-K–12 school food preparation and media analysis programs.

The current study's "unknowns" nurtured my flexible approach to shaping the course. For instance, I could not anticipate the teens' reaction or approach to designing their own course-related recipes and food preparation schedule; the intention was to allow the teens to move in what directions they would. Unexpectedly, about half of that first session (one hour out of two) consisted of encouraging these teens to select what recipes they wanted to cook, resulting in confusion, as it became increasingly apparent that they had rarely been asked what "they" wanted to do during previous formal or informal educational experiences (details about this struggle are shared and analyzed below). Christensen (2003) and Seabrook (1991) both saw this kind of reticence on the part of their groups of English and Social Studies students, respectively, as their students struggled with what should be done when teachers are not telling students what to do. This often requires teachers to "shake" any preoccupation with what students should be doing and to really think about "what [one] want[s] to teach and why it [i]s important" (Seabrook, 1991, p. 478). However, when you promote choice, flexibility, and shared power as part of your teaching goals with groups of students who are not familiar with this approach, it is important to build a transition from their familiar authoritative structure to an unfamiliar flexible, dialectical process of teaching and learning.

To let critical pedagogy drive this intervention, on the first day of the course students were invited to explore cookbooks and online recipe search engines to build not only the meals that they would prepare over the next four weeks, but also those that they might like to prepare at home in between course meetings. By exploring and choosing recipes as a group, these teens and I engaged in a "problem-posing" (Freire, 1970) educational approach to this initial step. Teens exercised choice selecting recipes and strategizing with me about important aspects of the food preparation process (for instance, timing of the recipes in relation to the two hours allotted for the course). The teens were positioned with the teacher through this participatory action research (PAR) approach, and the culmination of this activity was a democratic voting process wherein all teens voted on each recipe to see which ones would ultimately become a part of the course recipe book and possibly the series of meals.

Before the second session, I printed the recipes and placed slip-covered copies inside binders, dividing the recipes according to four categories: main courses; side dishes; snacks, breads, and appetizers; and desserts. Students then used magazine pictures or drawings to individualize their own recipe books. The students could elect to take the recipe book home each week or leave them at the community center. This promoted opportunities to share the recipes and food preparation methods with family and friends.

The course was always meant to be more than "just" a cooking class; media analysis was also a core mission. Throughout each of the remaining four cooking sessions, I integrated conversations about what students saw on television, in movies, and at school in relation to nutrition and food marketing, inviting them to question the messages. For instance, while making an apple pie from scratch, I half-jokingly asked how many commercials they'd recently seen for apples. I suggested the *unprocessed* apple is not fodder for a profitable ad campaign and would likely be viewed as "not marketable."

When mentioning this kind of food-related information to the teens, conversation ensued about what their families and communities might remember from past food experiences, as well as what kinds of foods they see advertised versus those overlooked by marketers. Also, I suggested that students try preparing snacks other than those that are prepackaged a few times in between course meetings, encouraging them to share stories about health-related as well as social experiences resulting from this additional food preparation process.

"Oh, the Boys Don't Want to Cook—They Just Want to Eat the Food": Disrupting Perceptions of Boys as Disinterested in Cooking

The female teens at the community center, from day one, would repeat the phrase heading this section: "Oh, the boys don't want to cook—they just want to eat the food." However, when speaking with the boys, largely on an individual basis, I heard a more complicated story. The boys did appear to have less experience cooking. On the first day of the class, a male teen, Keenan (around 13 years old and African American), told me that he'd tried to make a cake from scratch a few months before, and when his mother found him preparing it in the kitchen, she asked, "Why are you doing that? I've got a cake mix right here!" During the third week of class (and second week of cooking), one 19-year-old male student of color, Fenton, was present to stir the taco meat on the stove (no one below the age of 18 was allowed to use the stove at the community center). Fenton noted that he had never cooked anything before. I did not hear any of the girls or young women say they had no cooking experience; if anything, they noted that they already knew how to cook. Yet the male teens endlessly recounted stories of either being actively dissuaded from using their home kitchens or else having limited experience with cooking.

The girls felt more "at home" in the kitchen, seeking less direction from me than did the boys, and I noted that any co-ed groups formed at stations typically reflected the girls directing the boys to mix certain ingredients or to seek out certain cooking implements. I found that I'd made assumptions about basic skill sets that were largely mobilized by the girls but not the boys. For instance, during the third week of cooking, as a group of three 13- and 14-year-old male African Americans were preparing the cheese filling for a lasagna, one called to me, "Is it really one *cup* of oregano?" I responded that it should only be one *tablespoon*—at which point Alicia, the female staff member assisting me, flew over to the group to see what they were doing with the recipe. Reflecting on this moment, I realized the importance of teaching measuring information during the first class, including a "crib sheet" with their printed recipes and ensuring that all recipes followed the same measuring shorthand.

But if the boys started out less experienced at cooking, they did become actively involved and interested. One beneficial and unanticipated outcome was in relation to the salad. Alicia noted that she'd never seen the boys eat so much salad; only a small bowl of salad remained from the large one prepared with spinach, red leaf and iceberg lettuce during the second week of class. When considering this response from the boys, in particular, I speculated about the competitive element involved in consuming "the most" salad among the boys—but I also heard many of them commenting about the pride they were taking in helping to prepare the salad and other dishes for everyone to share. This sense of pride in ownership of the outcome as well as the activity of preparing the meals is something that I heard far more often from boys than girls.

Other evidence of the boys' growing interest in cooking came from their parents. While washing dishes following the third cooking class, Alicia noted that several parents (mostly of the male teens) had come in asking what the staff members were "doing" to their teens. One 13-year-old African American male, Benny, had prepared a cheesecake at his father's house; upon returning to his mother's house early the next morning, he entered her bedroom with a slice of the cake and said, "Try this!" She responded that it was early in the morning and couldn't she eat something else first. Benny responded, "No—you have to try this!" Now, it seems, Benny speaks of nothing but cooking and wants to learn how to prepare food and help his parents shop for it. Alicia reiterated that this was only one of many examples and that she was really happy that I was running this program—especially for the boys, who rarely had opportunities to learn how to cook.

Informal conversations with boys, girls, and female staff workers frequently developed while washing the mountain of dishes following each cooking adventure, and often this led to reflections on how food preparation experiences at home related to gender relations within the family. As noted, most stories revolved around the boys and their uncertain relationship with food preparation at home. As the class meetings continued, more stories like those related by Alicia about the male teen's newfound interest in food preparation began to surface. On the other hand, for female teens, these informal conversations became opportunities to discuss their academic and life aspirations, as well as pressures to conform to their parents' expectations. This signaled to me that this critical pedagogical approach to guiding the teens through a cooking course had opened a space for the boys to engage with and speak about food preparation, while the girls, who largely had an established comfort level with these processes, were able to stand on this common ground with me and speak about what else they wanted and could do with their lives.

We can see, then, how an after-school program could involve not only a consideration of basic cooking knowledge, but also a disruption of preconceived notions of masculinities. Issues of gender are never unproblematic, but they can be complicated in critical ways through critical pedagogical methods. Culture, race, religion, ethnicity, and social class—a blend of factors that often structure boys away from kitchen-related duties—however, confines girls within the food preparation context. And while the boys consistently sought my direction throughout the course, the girls did not do so as often, repeatedly noting how they prepared meals for themselves or other family members at home. As I continued observing the teens and pondering their comments and stories, I thought about how a school food program might be crafted so that it reflects consideration of these gendered practices.

Optional Ingredients: Reconceptualizing the Cooking Experience Through the Lenses of Religion and Food Allergies

During the second week of class—the first day of cooking—I brought many different salad ingredients for students to try, among them walnuts. Alicia whisked these away to the main office until the end of the class: one student, she said, was allergic to nuts. This was my introduction to food allergies at the community center.

Value systems surrounding food also existed affecting what many teens could taste and what recipes could be prepared each week. For the second week of cooking, I decided to bring ground pork instead of ground beef only to discover that some teens did not eat pork—whether for religious or cultural or personal reasons I did not know. A food policy ecology includes a consideration of the many cultural and religious perspectives influencing food consumption in society as well as the ways in which food allergies present barriers to what schools and community centers are able to feed children. These experiences highlighted burgeoning food allergies in the United States and the ways in which cultural and religious backgrounds influence relationships with food. Because of these very real food parameters, a highly structured atmosphere permeates the Prospect Clubhouse kitchen.

In future food preparation and food-related media analysis interventions, I might begin the course by clearly outlining my expectations for the teens—select recipes, create a cookbook and meal preparation schedule, prepare meals each week, and discuss food-related media issues—and providing them with specific examples of what recipes might be acceptable or what parameters might prevent some recipes from being considered, like excessive time for preparation, cost, or dietary restrictions. Discussing these additional considerations would provide opportunities for bringing the students' cultural backgrounds and relationships to food into the course.

Folding Stories into the Mix: Critical Pedagogical Approaches in a Highly Structured Atmosphere

On the first day of the class, I received quizzical looks from the teens when I initially asked them to list possible recipes for us to prepare. This was to be my introduction to the break from routine my cooking class would represent at the community center. This "break" was more of a negotiation between the rigid scheduling and social control associated with many activities at the center, and the open, flexible atmosphere I hoped to cultivate with students to ask questions and share stories throughout the food preparation process. There was clearly a tension between the center's approach and mine. I believe the Prospect Clubhouse responds with a high level of discipline and structure so that the children and teens "know what to expect." Clear rules and expectations can create a sense of safety, and over the five weeks during which I observed interactions at Prospect Clubhouse, I found that the children, teens and staff seemed very comfortable in this highly structured environment, with few of them pushing against these boundaries through words or actions.

For instance, before the first class meeting began, I watched as staff members directed 8- to 13-year-old children to line up quietly. Very few children spoke and all fell into place in a practiced manner. I observed this lunchtime line-up each week. A few times children were chastised for talking in line, adding one to two minutes to their wait with each remonstrance.

Throughout the first class meeting, I had the distinct impression that the staff members were gauging whether I could "handle" the group of teens. I suppose in honest reflection, they had reason to be wary; as I mentioned, I have never taught a cooking class, and I was a brand new volunteer at the center and an outsider to the community. So, during that first session, a female and male staff member intermittently entered the kitchen to inquire after what the students should be doing and whether they should be in the computer lab (some were gathering recipes), as well as outright chastising students and guiding them back into the kitchen. By the time I began the raised voice and then shouting round necessary to guide the students through the democratic process of selecting the "top five recipes" and the four meals to be prepared, staff members had all but disappeared, signaling that I'd won their confidence in my classroom management capabilities. Perhaps, the yelling did the trick.

Highly structured rules, codes, and regulations followed me into the community center's kitchen, though the staff never explained these before I began working with the teens, which only surfaced, it seemed, when I incurred an infraction. For instance, as alluded to above, there were age-related safety codes preventing children under 18 years of age from using the stove, which I was not told until a 15-year-old boy was helping stir the ground beef mixture to be used to make *pastelitos* during the second class meeting. I knew that such rules were there for legal health and safety reasons, but as an outsider, I was not immediately made aware of these regulations and only learned them gradually, while they were completely normative practices and expectations for all children, teens, and staff members. Every outsider to an organization has to learn the ways things are done in the new context, of course, but in this case it lent a level of surveillance to the tension already involved.

I probably should have started forming separate stations for small groups of students from the first day of cooking. The second week of cooking felt much easier because ten minutes prior to the start of class I arranged stations around the kitchen: one for preparing the ice cream cake, two for cleaning and chopping the vegetables for the taco salad, and one for prepping the macaroni and cheese and later the meat for the taco salad. Negotiate unfamiliar terrain was central to the successful completion of each week's meal, as well as developing levels of rapport with teens necessary to create opportunities for them to ask questions and reflect on their food-related experiences.

Overall this experience has shown me that any intervention into a food policy ecology should be flexible and that social equity aspects can be addressed with critical pedagogical approaches to food preparation. Since any ecology "is complex and subject to constraints that policymakers [and, I would add, educators] sometimes cannot overcome, control for, or even recognize" (Weaver-Hightower, 2008, p. 160), flexibility is a central requirement, and can be facilitated by critical pedagogical approaches and best analyzed by policy ecology methods.

Implications and Future Directions

The after-school food preparation and media analysis program seemed to trigger a desire in the teens to prepare food beyond the classroom, but I wonder how far it can flourish in the cases of the male teens, who are continuously *dis*couraged from entering the kitchen. Further, while I was able to introduce broader food purchasing and preparation issues and discussion of marketing ploys in the midst of conversation, any hope for group conversations went out the window while maintaining order and focus in a short class time. Still, a structured conversation apart from cooking or some other activity might encourage the teens to explore new possibilities in food, and media consumption. One way might be creating structured questions or assignments that would entail them returning home over the week and engaging in conversations with their family and friends, investigating television ads and shows, and even monitoring their fast food intake or that of others (McKinley et al., 2005). Doing so emphasizes the ownership aspect of the course that they appeared to thrive in, fostering a sense that they're detectives or researchers of their own lives, shifting their attention to those aspects of which they may not have been previously aware. Building on the informal conversations that occurred with small groups of teens in this study, the time during which food is cooking or being consumed by the teens would provide opportunities to encourage conversations about food-related media issues, while I lend structure in ways that might feel unstructured and relaxed for the teens.

Interventions like the one assessed in this paper have implications for the nutrition and food preparation education in North America, helping to break what has become generational dependence on fast food and pre-packaged food industries. Cooking skills and critical media skills might have positive impacts on this trend. Clearly more than a lack of skills is involved, including issues of access to nutritious food in urban food deserts (Raja, Ma, & Yadav, 2008) and cost, but cooking skills and media literacy are important facets nevertheless. In order to tease out these complex connections, future studies might consider integrating individual interviews with students to probe their broader educational experiences and existing knowledges about food.

Naturally, an after-school community program like this has differing concerns from a school program. Even for a school food policy ecology, though, one should consider these essential components in relation to generating flexible and socially just policies for food preparation among low-income students of color. Perhaps until the pressures of standardized testing and restricted budgets are at least partially alleviated in schools, an after-school food preparation and food-related media analysis intervention may be in the best position to flourish and be tracked by researchers in the hope of producing models to be used in schools. Future research might explore similar interventions in other countries with highly industrialized Western diets (i.e., Britain, Australia, Canada) and with different segments of their populations (i.e., in terms of race, ethnicity, social class, and gender). This might provide insight into bridging the school and after-school program gap in order to facilitate similar critical pedagogical approaches to nutrition and marketing awareness across educational institutions.

References

Albon, D., & Mukherji, P. (2008). *Food and health in early childhood.* London: Sage.

Baer Wilson, D., Musham, C., & McLellan, M. S. (2004). From mothers to daughters: Transgenerational food and diet communication in an underserved group. *Journal of Cultural Diversity, 11*(1): 12–17.

Berry, D., Davidson, M., & Grey, M. (2004). Preliminary testing of a program to prevent type 2 diabetes among high-risk youth. *The Journal of School Health, 74*(1): 10–15.

Bisset, S., & Potvin, L. (2007). Expanding our conceptualization of program implementation: Lessons from the genealogy of a school-based nutrition program. *Health Education Research, 22*(5): 737–746.

Burnet, D. L., Plaut, A. J., Ossowski, K., Ahmad, A., Quinn, M. T., Radovick, S., ... Chin, M. H. (2007). Community and family perspectives on addressing overweight in urban, African-American youth. *Journal of General Internal Medicine, 23*(2): 175–179. doi: 10.1007/s11606-007-0469-9.

Christensen, L. M. (2003). The politics of correction: How we can nurture students in their writing. *The Quarterly of the National Writing Project, 25*(4). Retrieved September 16, 2009, from *http://www.nwp.org/cs/public/print/resource/951?x-print_friendly=1*

Cunningham-Sabo, L., Bauer, M., Pareo, S., Phillips-Benally, S., Roanhorse, J., & Garcia, L. (2008). Qualitative investigation of factors contributing to effective nutrition education for Navajo families. *Maternal and Child Health Journal, 12*: S68–S75. doi: 10.1007/s10995-008-0333-5.

Epstein, L. H., Handley, E. A., Dearing, K. K., Cho, D. D., Roemmich, J. N., Paluch, R. A., ... Spring, B. (2006). Purchases of food in youth: Influence of price and income. *Association for Psychological Science, 17*(1): 82–89.

Fine, M., Torre, M. E., Boudin, K., Bowen, I., Clark, J., Hylton, D., . . . Upegui, D. (2004). Participatory action research: From within and beyond prison bars. In L. Weis & M. Fine (Eds.), *Working method: Research and social justice* (pp. 95–119). New York: Routledge.

Fontenot Molaison, E., Connell, C. L., Stuff, J. E., Yadrick, M. K. & Bogle, M. (2005). Influences on fruit and vegetable consumption by low-income black American adolescents. *Journal of Nutrition Education and Behavior, 37*: 246–251.

Foreman-Peck, L., & Murray, J. (2008). Action research and policy. *Journal of Philosophy of Education, 42*(S1): 145–163.

Freire, P. (1970). *Pedagogy of the oppressed.* New York: Continuum.

Freire, P. (2005). *Teachers as cultural workers.* MA: Westview Press.

Granner, M. L., Sargent, R. G., Calderone, K. S., Hussey, J. R., Evans, A. E., & Watkins, K. W. (2004). Factors of fruit and vegetable intake by race, gender, and age among young adolescents. *Journal of Nutrition Education and Behavior, 36*: 173–180.

hooks, b. (1994). *Teaching to transgress: Education as the practice of freedom.* New York: Routledge.

Izumi, B. T., Rostant, O. S., Moss, M. J., & Hamm, M. W. (2006). Results from the 2004 Michigan farm-to-school survey. *The Journal of School Health, 76*(5): 169–174.

Johnson, G. A. (1982). Organizational structure and scalar stress. In C. Renfrew, M. Rowlands, & B.A. Segraves-Whallon (Eds.), *Theory and explanation in archaeology* (pp. 389–421). New York: Academic Press.

Kincheloe, J. (2005). *Critical pedagogy.* New York: Peter Lang.

Kolbe, L. J., Kann, L., & Brener, N. D. (September 2001). Overview and summary of findings: School health policies and programs study 2000. *The Journal of School Health, 71*(7): 254–259.

Lalonde, C. (in press). Power personified: Graduate students negotiating Hollywood education. In C. Stephenson Malott & B. Porfilio (Eds.), *Critical pedagogy in the 21st century: A new generation of scholars.* Charlotte, NC: Information Age Publishing.

McKinley, M. C., Lowis, C., Robson, P. J., Wallace, J. M. W., Morrissey, M., Moran, A., & Livingstone, M. B. E. (2005). It's good to talk: Children's views on food and nutrition. *European Journal of Clinical Nutrition, 59*: 542–551.

Morgan, D. L. (1997). *Focus groups as qualitative research, 2nd edition.* Thousand Oaks: Sage.

Morgan, K., & Sonnino, R. (2008). *The school food revolution: Public food and the challenge of sustainable development.* London: Earthscan.

Mozaffarian, R. S., Wiecha, J. L., Roth, B. A., Nelson, T. F., Lee, R. M., & Gortmaker, S. L. (2010). Impact of an organizational intervention designed to improve snack and beverage quality in YMCA after-school programs. *American Journal of Public Health, 100*(5): 925–932.

Murray, S. (2007). *Moveable feasts: From ancient Rome to the 21st century, the incredible journeys of the food we eat.* New York: St. Martin's Press.

Nestle, M. (2007). *Food politics: How the food industry influences nutrition and health* (Revised ed.). Berkeley: University of California Press.

Petrini, C. (2007). *Slow food nation: Why our food should be good, clean, and fair.* New York: Rizzoli Ex Libris.

Pollan, M. (2008). *In defense of food: An eater's manifesto.* New York: The Penguin Press.

Raja, S., Born, B., & Kozlowski Russell, J. (2008). *Transforming food environments, building healthy communities.* Planning Advisory Service (PAS) series. Number 554. American Planning Association.

Raja, S., Ma, C., & Yadav, P. (2008). Beyond food deserts: Measuring and mapping racial disparities in neighborhood food environments. *Journal of Planning Education and Research, 27,* 469–482.

Rinderknecht, K., & Smith, C. (2004). Social cognitive theory in an after-school nutrition intervention for urban Native American youth. *Journal of Nutrition Education and Behavior, 36*: 298–304.

Roberts, P. (2008). *The end of food.* New York: Houghton Mifflin Company.

Rosecrans, A. M., Gittelsohn, J., Ho, L. S., Harris, S. B., Naqshbandi, M., & Sharma, S. (2008). Process evaluation of a multi-institutional community-based program for diabetes prevention among First Nations. *Health Education Research, 23*(2): 272–286.

Seabrook, G. (1991). Teachers and teaching: A teacher learns in the context of a social studies workshop. *Harvard Educational Review, 61*(4): 475–485.

Suarez-Balcazar, Y., Hellwig, M., Kouba, J., Redmond, L., Martinez, L., Block, D., ... Peterman, W. (2006). The making of an interdisciplinary partnership: The case of the Chicago Food System Collaborative. *American Journal of Community Psychology, 38*: 113–123. doi: 10.1007/s10464-006-9067-y.

Weaver-Hightower, M. B. (2008). An ecology metaphor for educational policy analysis: A call to complexity. *Educational Researcher, 37*(3): 153–167.

Weis, L., Kupper, M. M., Ciupak, Y., Stich, A., Jenkins, H., & Lalonde, C. (2010). Sociology of education in the United States, 1966–2008. In S. Tozer, B. P. Gallegos, A. Henry, M. B. Greiner, & P. G. Price (Eds.), *Handbook of research on the social foundations of education* (pp. 15–40). New York, NY: Lawrence Erlbaum Publishers.

• CHAPTER EIGHT •

Going Local
Burlington, Vermont's Farm-to-School Program

Doug Davis
Dana Hudson
Members of the Burlington School Food Project

The Burlington School Food Project serves as a model for bringing local farmers and the food they produce into school districts and summer meal programs. Through collaboration with farmers, school staff, students, and community volunteers, local farmers expand their market and more students participate in the meals program. This is a beaming example of Vermonters' ability and commitment to working together and achieving success. I hope that we can begin to replicate this project around the state and across the country.

–Senator Patrick Leahy, 2010

It was the last Friday of March around eight in the morning before the big event. None of us knew how this would play out. Our team had been planning for months, and it was finally time to set it all up. We walked into the giant, dark exposition hall and looked around at the vast empty space. As is the case with all that we do, we just rolled up our sleeves and went to work. Volunteers began to arrive and the room took shape. Could we really pull this off? At this point, having spent over 30 winters in Vermont, I was just relieved that we were not in the midst of a spring blizzard. Maybe this crazy idea was going to work after all.

The idea for the Vermont Junior Iron Chef Competition came out of a Burlington School Food Project meeting back in the spring of 2007 as a potential fundraising idea. This would be a cooperative effort between Burlington School Food Project and Vermont Food Education Every Day. I was skeptical: who would want to come to such an event and why? We had been seeing a lot of success in working with farmers and students to reshape our menus, procedures, and practices, but the thought of a state-wide event to promote this change seemed daunting. Cooking shows and cooking competitions were becoming really popular, and one of our local food celebrities had just been featured on one of the shows. But we are not television producers, and to me this seemed like a huge undertaking.

The planning was intensive and the time commitment insane. Prior to taking this on the Burlington Food Council and Burlington School Food Project took up about six to eight hours a month. This time was spent reporting our progress to partners and brainstorming ways to incorporate more local foods and nutrition education into our programs. This competition was a new beast. We needed to build it from the ground up: create the rules, find and secure the location, get other schools involved, advertise, sell sponsorships, and one of the most important steps—we had to get our superintendent to sign off on the original outlay of time and resources. Fortunately, our superintendent, Jeanne Collins, saw the importance of what we were trying to accomplish and supported us through every step. She saw this event as a way to spread the good work being done in Burlington to many other school communities throughout the state, so she encouraged us to proceed. Her management style is to work with and support her directors and to do all that is necessary to have all district programs excel. It sounds like that should be the norm, but from my experience in the education world, most superintendents are afraid to allow that much latitude to their directors. Most would rather say no to something new than to risk blame and failure. It is easy for me to say that without her confidence and support, this event and many of our other initiatives would look substantially different today.

Pat Matton, Burlington School District Food Service employee, Abbie Nelson of Northeast Organic Farming Association/Food Education Every Day, and Bonnie Acker, parent, worked tirelessly to set up committees to oversee every possible aspect of the event. The entire process was done manually, on paper. As is the case with much of our farm-to-school initiatives, we were blazing new trails and had to learn as we went along. All the while, we had to maintain our day-to-day business of creating 5,500 school meals and handling all of the challenges that went along with that.

The first annual event was a success: over 50 teams competed; it received great media coverage; lots of local food was consumed; and most importantly, there were lots of happy, engaged and excited students. One of the original winners of the event was the Healthy City Youth Farm team. This was a group of Burlington students that had been involved during the previous summer with the Healthy City Youth Farm program, which works to engage "at risk" students in farm and garden activities. The director of this program, Jenn McGowan, has been an active member of the Burlington School Food Project from the beginning and has done so much to engage her kids in many of our initiatives. Having a team in the first Junior Iron Chef competition seemed like a great fit. Her team of kids was intimidated at first; they are not kids that are used to being competitive and certainly not used to being put under pressure to perform. They were competing in the high school competition against teams from vocational culinary programs, all dressed in chef whites. But they worked hard, kept their focus, and had a blast. They were so excited to get that Best in Show award. More importantly, one of the team members took one of the donated scholarship prizes and is attending culinary school, making her the first person in her family to go to college.

The event brought in over $10,000 to the Burlington School Food Project and Vermont Food Education Every Day. The proceeds were used to fund another initiative, the Farm-2-Arts card fundraiser (see below). Teams left the first event asking when the next one would be. We had created a monster! Even with all the success, we knew that if we were to do it again, much would have to be streamlined. Though I may have voted to call it quits, not doing another one was not an option. Sometimes, you have to "be careful what you wish for" in farm-to-school programs, because things often come true.

The second Junior Iron Chef went better than the first. We engaged more students and farmers, continued to help create new partnerships and programs, and made more money. This time we used the funds to hire a local web site design firm, and we created the Vermont Junior Iron Chef Website (www.jrironchefvt.org). This was a giant step forward. Most applications and communications were now done online. Rules, dates, and procedures were available 24/7. Also a big help, Vermont Food Education Every Day funded a part-time position to support the Junior Iron Chef. Now the event had become a workable, streamlined process. In the planning for the 3rd Annual Junior Iron Chef Competition, we were more organized and more confident. Gathering teams, sponsors, and prizes was easier, and we were able to get our highest profile guests.

Lisa Pino was one of those guests. Lisa, the Deputy Administrator of the Supplemental Nutrition Assistance Program (the federal food assistance program formerly called "food stamps"), agreed to be a judge and our kick-off speaker. Lisa Pino is one of the most amazing people I have met in my years of child nutrition. She is empathetic, hard working, genuine, extremely intelligent and just an all-around great person. Lisa first came to Burlington in July of 2009 to view one of our summer meal sites with other dignitaries. At one point during that visit, we all looked around and could not find Lisa. She had sat down with a group of students and just started talking to them. I remember her saying something like "I work in Washington, DC, for President Obama, and he wants to know how things are going in Vermont." It was amazing: the kids immediately engaged with her, creating a moment for them that they will likely never forget. Then on March 27, 2010, Lisa Pino returned to Vermont to kick off the 3rd Annual Vermont Junior Iron Chef Competition and before ringing the honorary cowbell she got the students fired up with:

Jr. Iron Chef Is Farm Fresh!
Farm-to-School Is Super Cool!
And Local Rules!

Why *This* Work and *Why* It Works

By itself, Farm-to-School is a simple idea: increase the local foods in the schools while giving the students some hands-on food education experiences. But the implementation is not simple. When you have to incorporate this simple idea into the highly complex and broken system of school food, it becomes unimaginably difficult. Not impossible, as we'll describe below, but difficult.

Incorporating farm-to-school ideas into schools and school cafeterias is as complex as addressing hunger and poverty, because it is a generational issue. In the 1970s child nutrition programs changed. The reimbursement model allowed for competition for students' lunch money, which had schools competing within their own walls against vending machines, fast foods, a la carte options, and school store snacks for the same student dollar. Perceptions of "poor" kids changed when they were getting in line for the standard lunch meal, while the "rich" kids got to choose from the a la carte options that were usually less healthy but started to create a culture of "cooler" foods. This created a generation that does not see the importance of eating fresh, whole foods and decreased the respect for the art of growing food, the pleasures of cooking, and the importance of mealtime.

During these last 30 years, schools also have reduced the ability of school kitchens to create meals. The school cafeteria workforce has adapted to a brown-and-serve mentality and does not know how to deal with whole fresh foods. School nutrition programs have hence become the brunt of what is wrong with our food system: iconic hairnet-wearing lunch ladies serving mystery meat. The poorest and neediest have had to participate in this program that is under constant ridicule within the mass media and pop culture. Adam Sandler and Chris Farley singing and performing "Lunch Lady Land" epitomize this stereotype.

In the door comes Farm-to-School. At first glance it seems to be a fringe effort to connect a small population of our community—our farmers—to another market to peddle their products. However, a second glance reveals farm-to-school programs to be a multi-objective and multi-sector strategy to address farm viability, childhood nutrition, food access, economic development, community engagement, educational standards, and environmental impacts. This inter-sector synergy makes Farm-to-School a win-win for all involved and creates very few, if any, antagonists.

What has happened in Burlington in the last 7 years has truly shifted the cultural expectations and norms of what is eaten in the school cafeteria, and even in the households and restaurants in Burlington. Instead of focusing on what is wrong with the system and the school meal program, the effort has concentrated on working within the system and creatively developing new approaches and strategies. A model has been created that helps the students get the nourishment they need, involves the community to successfully contribute to its own food system, and improves the access to fresh and whole foods from local farms.

Overall, we have figured out how to get fresh, local, whole products to all the children eating meals in the Burlington School System. We are definitely not done, but we have created the system to make it happen. By focusing on the process and not the actual product, we have created the vehicle for change. One farmer that sells to the Burlington schools, who also has children in the district, sells most of his products to high-end restaurants. But he was thrilled to start working with the schools to find a price point for products that the schools could afford and he could sustain so that his children and their friends and their friends' friends could finally have access to the food and messages farm-to-school programs convey in a way that was not previously available. His food is now in the forefront of school food system change, on the student lunch trays and school salad bars, changing things from inside the system outward. That is what Farm-to-School is all about: creating innovative new systems within the current system.

Just getting more fresh fruits and vegetables on a school meal tray and consumed by students is not Farm-to-School. It is the process of getting the cafeteria, the classroom, and the community to come together to benefit and support a cultural shift in what is expected and what is feasible. Just serving better food is not going to make everything all right. Figuring out how everyone benefits—the farmer, the school, the community, and most importantly the students—is the measurement of success.

The History of the Burlington School Food Project

We are the first to admit that Burlington is not like everywhere else. It has been found to be the healthiest city in the United States, with a strong history of neighbors helping neighbors. Vermont has the most farmers' markets per capita and the most community supported agriculture programs per capita. We have had 10 successful years of a "buy Vermont first" campaign that is not necessarily about food, but about all goods. Seen as an agricultural state, Vermont has a much-respected seal of quality on its exportable food goods. We just don't give in. We don't have billboards or Target. We have the only state capital that doesn't have a McDonald's within the city limits, because the citizens kept it out. Overall, the culture and landscape is right for Farm-to-School to flourish and succeed.

The planning for the Burlington School Food Project began in 2002 when at a Burlington Legacy town meeting food was a number one concern that came out of the citizen surveys. From this instigation, the city, the school district, and a number of non-profits sat down together to discuss and plan a way to address this concern and make some needed changes. The main intent was to create big change, but to let it grow slowly, sustainably, and with complete support and participation from all sectors of the community.

Vermont Food Education Every Day, a Farm-to-School partnership project of three Vermont non-profits—Food Works, the Northeast Organic Farming Association of Vermont, and Shelburne Farms—had been working with rural Vermont schools for 3 years on initiating Farm-to-School ideas. They were ready to dig deeper within a community and pilot a longer-term relationship to build the community support that would be able to sustain any changes. So Vermont Food Education Every Day, in partnership with the school district, the Intervale (a nationally recognized center for sustainable agriculture), and a number of other key partners, applied for and received a United States Department of Agriculture Community Food Project Grant. Since it was only partially funded, we re-evaluated the project goals and activities to get the most and the furthest reaching impact.

The Burlington Food Council was an initial and key element to launch this project. By engaging community members in an organized and regularly scheduled meeting that was focused on empowering them in a working group format, community ownership was established. This venue allowed for education on everyone's part. People learned to walk in each other's shoes. The parent who once was the biggest naysayer about school food became the biggest supporter of the food service director once she learned the barriers and constraints to making a school meal program solvent. And the food service director, who wanted to lock the door when the activists came knocking, learned that there was great opportunity and great power in numbers if everyone could learn to create win-win solutions together.

In its first year, the Burlington Food Council developed and implemented a Community Food Assessment, creating a baseline of information to help decide what and how to build the Burlington School Food Project. Parents, teachers, food service workers, farmers, businesses, school administrators, and every 4th and 7th grader in the city were surveyed. Information about transportation, education, food access, demographics, and other pertinent information was collected. All this was compiled into an assessment report that laid the framework for what were the opinions, behaviors, and current practices around food for youth in K–12 schools. During this first year, the Burlington Food Council also developed an initial action plan to launch the project.

From this action plan stemmed numerous educational activities including teacher professional development, farm-based field trips, food-based lessons integrated into teaching standards, after-school programming, and summer educational offerings. The plan also put in motion infrastructure and systems changes such as winter farmer and food service planning meetings, food service staff training, eating environment alterations and additions, and vendor discussions.

One of the most important activities that emerged from the action plan has been the development of running taste tests for students throughout the school year. Before any new food item is introduced on the lunch line or in the breakfast program, students first have a taste test. Sometimes these taste tests actually have student participation in the harvesting, preparation, cooking, and serving of the sample. Students also are involved in surveying each other about their willingness to try these new foods and whether they liked them and would eat them again. Most recipes go through multiple iterations before a final product is incorporated into the meal program. Many things are taken into account before an item makes it into lunch including student preferences and palate, ethnically appropriate ingredients for our diverse student populations, food and personnel costs for production, ability to replicate the recipe beyond the initial offering, seasonality and/or year-round access to ingredients, equipment, time and training, and transferability to all schools in the district.

Initially after each taste test, there was a time to reflect and evaluate success or failure. Ongoing evaluation and reflection throughout the project created a culture of open learning and dialogue that strengthened the commitment between project partners and food council members. Out of this structure and history bloomed the supportive environment of public and private partnership between city, school district, farmers, food businesses, nonprofits, and citizen activists. Egos are left at the door, and all ideas that are implemented are intended to meet everyone's interests and agendas. This multi-objective strategy makes Farm-to-School in Burlington a win-win for everyone involved. This also has created the environment for the Junior Iron Chef event and the following projects to grow successfully. These stories, shared by different partners involved in the Burlington School Food Project, are only some of the initiatives under way in Burlington since 2003.

Stories from the Field and the Partners that Make It Happen

100,000 Stories

By Bonnie Acker, Founding Food Council Member, Parent, Artist, and Community Activist

I write notes. I know it's old-fashioned, but I love to anyway. I also love to strike up conversations. Last year at the post office, I got into such a lively exchange about school-food change with one of the clerks that I forgot to put a stamp on a letter. Later, one of the counter staff, Stacey, called to tell me that another staff person, Caroline, had added a stamp.

The next day, I returned to pay for the stamp, thank the wonderful clerks, and give each a pack of our note cards. A gal in line noticed the cards, asked about our project and then exclaimed, "I can help!" Within a few weeks, Susan had helped design and sew beautiful banners for our celebratory Junior Iron Chef Vermont event. Discovering supporters doesn't get any better than that!

So I'll back up a bit. A bunch of us in 2004 had started to focus on school food in Burlington. We needed—and still need—additional funds for more whole, local, and fresh foods. One afternoon, Doug, the head of our school-food program, reflected, "Bonnie, you know my wife Robyn writes a lot of cards. She buys a lot of cards. Could we produce some as a fundraiser?"

"Schools could sell them instead of wrapping paper and candy!" I beamed. "Less chocolate and more kale, that's my mantra for organizing. Let's do it!"

Years earlier, while my young daughter was enjoying a morning at Audubon day camp, I had been sitting in a meadow sketching daylilies with pastels. Soon after, I donated the scenes for the group's upcoming auction, and the executive director Bill exclaimed, "One must be a poster, and I have a printer in mind!" It was KBA near Burlington, the oldest press manufacturer in the world. Showcasing their world-class equipment they donated work, and a large Audubon Center poster was soon off the press.

Fast-forward seven years; we were dreaming of note cards. I remembered KBA and called up. After hearing about our accomplishments, Eric, the vice president of marketing, replied, "We'll do anything you want." A miracle. For the next four years, I called KBA every six months. Terry, the receptionist, made reaching people a joy. I explained to the pre-press manager, Jeff, that we were waiting until we felt worthy of their offer. "Anything," he reiterated, "whenever you are ready."

Finally we felt real momentum. Nearby farmers were growing for our salad bars, food-service staff were creating more dishes from scratch, local pizza-makers were on board, teachers were excited about farms, and students were really enjoying the changes. Early in 2008, we told KBA we were ready.

It was time to look for images to give a face to our stories. Katharine, a painter and supporter and mom of kids in the Burlington schools, picked up the phone. "Could you donate the use of a couple images?" Magically, eight other Vermont artists—in addition to Katharine and myself—immediately agreed. "I am thrilled to be part of your wonderful project," one wrote. "It is work dear to me."

We also needed a business plan. Seeking advice from Burlington's Community and Economic Development Office, a kind staff person praised our enthusiasm, observed our competition and advised us against note cards. We were undeterred. We felt we had no choice, for there were over 2,500 students depending on our school meals.

Pat, Doug's co-worker in our school-food program, started to organize meetings and coordinate tasks. We came up with a visionary name, Vermont Farm 2 School Arts, and a catchy logo. Abbie, one of Vermont Food Education Every Day's statewide organizers, wrote card-back copy. Megan and Tom from Shelburne Farms figured out how their nonprofit could obtain trademark protection and handle the finances. We projected expenses and income. We were so confident! Yet we completely underestimated the effort it would take to sell our cards. KBA stayed with us. Rob offered to help with all pre-press tasks. Press-people Bruce and Paul suggested Astrolite paper from Monadnock Mills. "It's beautiful, all recycled, all post-consumer content." Crystal advised, "To go green, you have to spend green." Chris encouraged, "Call Monadnock in New Hampshire. I bet you could get a donation."

Heart pounding, I reached a receptionist who said a donation was impossible. But after hearing about farmers donating butternut squash and students painting harvest scenes and food-service staff bravely preparing new dishes, Cindy said she'd talk to the CEO. A half-hour later, she'd arranged for Monadnock to donate most of the paper for the cards and envelopes. Another complete miracle.

Within six months, tons of 24 x 40-inch made-to-order paper sheets had been shipped to KBA, set-up was completed, and the press run was scheduled. Chris had offered to produce a million cards but we had asked for less, envisioning twenty-five years of collaboration. In the fall, our 100,000 beautiful note cards came off a $4,000,000 press. Barely a month later, KBA's management decided to move to Texas to facilitate the shipping of press parts around North America. With no further assistance possible, we treasured what KBA had given us beyond words.

Just after New Year's in 2009, with the cards cut, scored and folded, envelopes and clear sleeves obtained, we were ready to assemble! Around long tables, wonderful City Market working members spent hundreds of hours producing 20,000 packs of five cards, five envelopes, and an insert. Stories of land trusts, jobs, Obama, Public Radio, children, and favorite recipes were shared, as friendships were renewed and many born. Bobby, our new school-district farm-to-school coordinator, salted away a hundred boxes full of promise.

Distribution, over the last year and a half, has been our joy and our challenge. Thankfully, people have loved our cards. A couple dozen schools have held successful fundraisers, easily selling 5-card packs for $10, keeping $4 for their own programs and returning $6 to us for our program here in Burlington and statewide Farm 2 School work. But word-of-mouth outreach, to schools and other groups and to individuals, has proved wholly inadequate. So we're finally developing a website. We've established retail outlets at City Market and elsewhere. We're bringing cards to conferences, and considering magazine ads. Sales of single $2.25 cards are outstripping sales of five-card $9.99 packs, so we're re-packaging thousands of cards.

We've covered our $15,000 of expenses and now have funds to buy local apples, bread and cheese, and hold community dinners where students, parents and the wider community can enjoy the delicious foods being served in our cafeterias. Our cards are beginning to make a difference. Doug's wife, Robyn, now has lots of cards. And many other people, using our cards, have shared thoughts about birth and death, gardens, soccer, birthdays, holidays, their many hopes and myriad dreams.

Will we raise the $100,000 that we were dreaming of a couple years ago? Only time will tell. We'll remain unshakable optimists, for the sake of so many in our community.

New Thinking Creates New Products

By Doug Davis

Much of our Farm-to-School success has not come from scrapping current models and starting over, but from thinking about what we do each day and just looking at it through a new lens. Two good examples are the Burlington School District Bread CSA and our partnership with Misty Knoll Farms.

Artisan bread. In the spring of 2009 I was discussing with Bobby Young, our Farm 2 School coordinator, the possibility of bringing in locally made organic artisan bread to our schools. I knew that cost would be an obstacle, but I didn't want it to become a barrier. The National School Lunch Program regulations require that a bread component be offered with every lunch served. We have switched to offering only whole grain bread products, and that component usually costs around 12 to 16 cents per serving, usually in the form of a whole wheat bread, burger bun or dinner roll.

I knew that our allocated cost per serving paled in comparison to the cost of production necessary to create the bread I wanted to offer my students. In order to be successful, we would need to create a new system to generate the revenue needed to offer this product. Bread has always played a major roll in history and culture. Many students in our school are low income and many are new Americans, so it was important to me to offer something more than a whole wheat burger bun (though compared to the white buns we were serving five years ago, the wheat buns are great). I spent a lot of time trying to figure out what we might be able to do. Finally, I had an idea. We could create a bread that would mirror the typical vegetable Community Supported Agriculture shares—in which the consumer pays a set amount before the season begins to help off-set the farmer's costs in return for a weekly share of the harvest that people are accustomed to buying.

My thought was to see if the Red Hen Bakery would be interested in being involved with such an idea. I thought we could offer these bread shares in ten-week increments to Burlington School District employees and have the loaves of bread delivered to their schools once a week. Based on the costs, it seemed that we would need to come up with loaves that the bakery could sell to us for around $2 each that I would sell to Community Supported Agriculture members for $4. We would then return the additional $2/loaf to Red Hen to purchase bread for student sandwich bars. If they agreed, we would need to get about 100 subscribers involved to make it work.

After an amazing tour of the bakery and hearing the great stories about the delivery of the huge ovens, we sat down with Randy and Harrison at Red

Hen to discuss the possibilities. I think that there was some rightful hesitance on the part of the bakery. This was pretty new to all of us. Sure, Red Hen was already selling some bread to local schools, but this was a whole new animal. We were trying to create in essence a whole new storefront for their products and a whole new generation of customers. There was no question that their bread was amazing, with many varieties to choose from. The question was: could we get enough members and facilitate the program and the distribution successfully? The last thing I ever want to do is create an unsustainable model. I did not want to imagine rolling this out, only to retract it several months later because it was unaffordable.

After about an hour or so of discussion, we decided to give it a try. We began working on a sales brochure and distribution model. In September of 2008, we launched the idea in three schools. The plan was to sell the 100 shares at $40 each for a 10-week term. This would gross the program $4000 and allow us $2000 to pay for student bread after the Community Supported Agriculture bread was paid for. We did not quite hit 100 members. We were closer to 85, but we thought we should give it a try anyway. There were some distribution challenges on our side and the paperwork was a bit more time consuming than anticipated, but overall the first session was deemed successful. With the CSA proceeds, we began buying sliced day old bread from Red Hen and purchased two sandwich presses to create grilled sandwiches to order daily. The addition of the grilled sandwiches really pushed the program into a higher gear. Teachers and staff that had chosen to participate as members now saw the bread being offered to students and adults for lunch. Students who may have never been able to try artisan bread now had it available daily. Other adults heard the Community Supported Agriculture bread story, and we were able to expand the number of schools where memberships were sold. During the 2009–2010 school year, we completed four Community Supported Agriculture bread sessions. This produced new customers for our schools and Red Hen while netting our program over $2500.

This relationship also has helped us open new relationships with other local bakeries. We now receive bread from Klinger's Bakery for creating our own croutons, and they are currently working with us to create a breakfast bar that will contain some local ingredients that can be served for our breakfast in the classroom program. We also worked with O Bread bakery to create a cheddar cheese roll using Shelburne Farms' two-year-old Cheddar Cheese (the best cheddar on the planet) in their amazing artisan bread. This cheddar roll (which we hope to retail), when served with a Vermont apple and milk, not only provides all of the federal requirements for a reimbursable breakfast, but also allows us to offer a truly local Vermont breakfast every week in every elementary classroom in Burlington.

This year we hope to expand the Community Supported Agriculture offerings we provide to include root crops, cheese, eggs, and possibly beef, pork and poultry. The important thing for us to remember is that we can never stop trying to expand on the local connections with our farmers and our community members because school lunch is not just for kids anymore. We are working to create future consumers who need to understand and play an active role in their own local food system, as well as the next generation of voters who will have the ability to support and protect our local agriculture, farmers, and environment.

Local chicken. Misty Knoll Farm is a free-range poultry farm about 25 miles south of Burlington in New Haven, Vermont. I began buying turkeys from Misty Knoll in 1988 for a restaurant I was managing at that time, and we have been buying turkeys from them for our home holiday meals for as long as I can remember. On my way home from work one day in the winter of 2008, my wife called and asked me to pick up something for dinner. Having already bypassed the large supermarket chains, I stopped into one of our local markets. I wasn't sure what to pick up, but as I often do, I just started wandering around the store looking at all of the local products that were available. I found myself at the meat case and noticed the Misty Knoll Chicken. There I saw whole birds, boneless and bone-in breasts, and thighs. Having spent time at Misty Knoll and having raised chickens myself, I knew that each bird had two drumsticks. I also knew that every time we cooked or bought a whole chicken or turkey to serve at home, the competition between our four kids was on to see who got to have the drumsticks. I began to wonder where all those legs were.

The next day I called Rob Litch at Misty Knoll and started talking to him about what I had seen. I told him that we offered chicken at least once a week and kids ate it pretty well. We were (and still are) offering a low fat whole grain chicken patty and/or nugget, but would much rather be offering a whole, fresh and local product if possible. Rob said that there wasn't much of a market for the drums and that the cost to debone them and use them in something else was not cost effective. He also said that due to the high cost of diesel fuel (about $5/gallon at that time), the Food Shelf couldn't afford to come pick the product up and he couldn't afford to drive the donation down himself. He asked me what I was currently paying for chicken and how much I used each time it was on the menu. Before I answered, I wanted him to understand my philosophy regarding our work with local farms and farmers. I explained that my overall cost of food for lunch needs to be around $1.15 and that I need to offer milk and a full salad bar daily. This leaves me about 45 to 55 cents to spend on the typical center of the plate item offered. I also explained that the goal of the Burlington School Food Project is to create long-term, mutually beneficial relationships with farmers, not to get products so inexpensively that the farmer cannot continue to do business with us or will only work with us until a higher paying customer comes along. As is the case with all of the farmers I work with, Rob is a bright guy and he saw this as a win-win situation. He knew that if I advertised Misty Knoll Farm products on my menus, the parents would see it, and he knew that he would benefit from their excitement. He also knew that I could take about 2,000 drumsticks at a time and he could do a single stop in Burlington on an already established delivery route. We agreed on a price per unit that is consistent with what I was paying for the processed product, and now I am able to offer his chicken on our school menus.

My goal for the 2010–2011 school year is to replace the patty and nugget offerings with whole chicken, and I hope that Rob will continue to play a large part in that work. One of the challenges in working with fresh raw chicken as opposed to cooked processed chicken is the potential for food-borne illness and cross contamination. In my twelve schools the level of infrastructure in my kitchens varies greatly. Many of the schools were built originally as "neighborhood walking schools" where students would walk home for lunch. As the need for school-based meals increased, kitchens were added. Several schools added them into storage areas or large closets, and some still lack the basic tools needed to create the meals my parents expect. Also, we feel that all students in elementary school should receive substantially the same meal regardless of which Burlington school they attend. As food safety is our number one concern, we looked at ways to make working with fresh meat products safer. I contacted a local processor who was producing several frozen ground beef items for our schools and asked him if I could have the chicken drumsticks delivered to him where he could cook them, test them, freeze them and ship them to my sites. He agreed to give it a try and to do it at a price that kept the product cost effective. Now my schools receive the cooked drumsticks in an oven-safe bag and all they have to do is heat them to the proper temperature. Sure, some kids still prefer the nuggets and patties, but this process and our ability to offer Misty Knoll chicken on a regular basis will likely cause more students to try and eventually choose this product over the processed one.

So, what is the moral of the story for me? Use every opportunity to see what is and what isn't available in your local markets. Sometimes it is what you don't see that holds all of the possibilities for change. And remember, if your spouse or partner asks that you stop and pick up dinner, recognize that you might be taking the first step on a really cool Farm-to-School journey.

City Market Member Workers and the Role of Community in Child Nutrition

By Bobby Young, Farm 2 School Coordinator

Our downtown grocery store in Burlington, City Market/Onion River Cooperative, has had great success in recent years and is seeing membership boom annually. They not only play a critical role in providing high quality, locally produced foods to consumers, they also are vital in transferring and retaining wealth to and in our city with every dollar spent in their store. Their main goal is food access and breaking down the barrier between high quality and high cost for many. They offer many programs to achieve their mission, such as their eating on a budget course, food-for-all program, and their member worker program. As a member of the Coop, you have the choice to become a member worker, where 2–4 hours of community volunteering every month gets you a hefty discount on your groceries. This includes helping out in the store, as well as being encouraged to get out and help various non-profits in the city, including the Burlington Schools Food Service and more specifically the schools' Farm-to-School efforts. As a main partner in the Burlington School Food Project, City Market has been a blessing in moving forward the transformation of school meals in Burlington and has brought together neighbors from all walks of life (mainly the ones who keep coming back to use a sharp knife in a hot kitchen) for lively discussions and vastly important work.

Just about everything you think could never happen becomes reality as community builds and neighbors come together for a commonly understood benefit for all. My first lesson in this came in the fall of 2008, when my former boss, Andy Jones, who manages the Intervale Community Farm, gave a call to say the farm had a ton of perfectly usable butternut squash that had been left in the field after harvest because it wasn't up to storage grade. The schools, he said, were welcome to take it for free. I'll have to say the word "ton" gets thrown around a lot in our daily language and really takes away from what a ton actually is. In the case of butternut squash, a ton is a pickup truck full of an awkwardly shaped staple of a Vermont storage crop. Not appreciating the true nature of what laid on the river bottom soils of the Intervale, I readily agreed to accept the donation, and the process began.

At the time, the Healthy City Youth Farm was still operating within the Intervale, and they quickly came to the rescue as determined harvesters, packers, and distributors of the squash. They even made it into the newspaper for their heroic efforts to transfer nutrition from the field to the lunch tray! So their truck pulled up to the high school's loading dock (probably more than once) and into the kitchens went the squash. Luckily, I realized we would need help to process the squash into a usable form, which for us meant cooked and mashed to be stored frozen. After several hours of processing, the finished product fit nicely into our already established process of making breakfast breads and soups, and our students enjoyed this locally grown product throughout the school year. Some of this product even made it into our Jr. Vermont Iron Chef competition as one of the featured local ingredients.

My caution around soliciting volunteer help to get this job done did not come from being timid or overconfident in my own capacity. It came from my realization and belief that volunteer efforts are not sustainable when a culture based on participation doesn't exist. How could this be, that in such an active city and school community, this great cause could go underserved? Our first reaction was an offensive against the community, but soon enough we realized it was not them, it was us. The relationship between the community and the school's food service just had not fully matured yet, and there was no platform in place to foster the growth of that relationship. So we took a hard look at how we could possibly get the help we needed to move forward. With over 5,000 meals to serve each day, we knew we just didn't have the resources to continue processing food from scratch and getting it on the tray.

Our first and honestly only place we knew best to focus was the developing opportunity to tap into City Market's member worker program. We were using it to our advantage all along, but we had never developed the common ground for it to flourish between the members and the school. Frequent and continual tasks were what was needed, and a set schedule for interested members to fall back on was essential. This is tricky when your operations are not really streamlined or developed enough to guarantee member workers 40 hours of labor every week. It became quite the juggling act trying to move production capacity in the kitchen forward while at the same time trying to garner and harness the energy and interest of member workers. After the initial long and brutal road of coordination, we have seen some success in the prep space, on the schools' salad bars and, most importantly, on the faces of the folks that helped make it happen.

In 2009, City Market member workers helped process 300–400 pounds of fresh produce in a single day every week to be distributed to and eaten from the district salad bars! The member workers gave the extra labor capacity that food service needed to be able to experiment with and implement new production lines to take the operation to new heights. After winter break, fresh salad bar prep transferred to our own staff, once the realization of possibilities occurred. With their help, we then moved on to another experiment, processing root vegetables and turning them into roasted roots, which has become a strong competitor to fries at the high school, and a replacement for fries in all schools grades K–8. Twice a week, 150 pounds of roasted roots are eaten by high school students, who would normally turn away if fried food was not on the lunch line. Not only has business been maintained, but nutrition has improved and our community connections have strengthened. Marketing, production and financial solvency are wrapped up in one package. We even have school board members now helping out as member workers. And the best part is that new friends and new experiences are made along the way.

Nature's Candy

By Sarah Kadden, Barnes Sustainability Academy, Shelburne Farms

"Its nature's candy!" quickly became the rallying cry of Mrs. Brown's first grade, as they buzzed between rows of crops at the Intervale Community Farm with farmer Andy Jones.

It is a cry we hear every time we work with students on farms at harvest time. The taste of fall produce picked with their own hands can only be compared to the sweetest confections. And it is often a surprise for students who do not have their own gardens to taste how sweet fruits and vegetables can be without the assistance of ranch dressing or caramel dip.

It was cabbage offered from the blade of a cleaver by Andy that got the most candy comparisons that day. The previous morning at Digger's Mirth Collective Farm, it was fennel, a taste fourth and fifth grade students loved so much that they convinced their teacher (with a little prodding from farmer Dylan Zeitlyn) to let them make fennel the subject of their upcoming persuasive writing assignment. They planned to lobby Food Service Director Doug Davis to include fennel in the following year's purchasing agreement. As the students wondered aloud if, in fact, they could make fennel into candy, a clever fourth grader named Kai had a better idea. "Guys," he said, "why don't we just eat it like this. We don't have to wash it, and, besides, then we get to use a knife."

Dylan, Andy, and fellow farmer Emily Merrell have become the beating heart of farm-to-school education in Burlington, Vermont. Dylan is a fixture at the annual Farmer Forum, where all the farmers who sell to Burlington's School Food Project meet, and all three are regulars at the Sustainability Academy at Lawrence Barnes School, where each are parents to children in kindergarten through fifth grade.

On any given day, Emily, Andy, Dylan, and Isha Abdi, another farmer/parent, can be found dropping off or picking up their own children and various friends. And almost as frequently, Emily is meeting with parents and staff on the school's wellness committee, Dylan is transplanting raspberries in the schoolyard, Andy is eating lunch in the cafeteria, and Isha is turning soil in one of the school's gardens with first graders. Any or all of them are frequent cafeteria and classroom guests, often leading taste tests or cooking.

These farmers have opened up their farms to students. They host field trips and welcome children into their workplaces on a regular basis. In the fields, students plant, harvest, cultivate, weed, play, smell, taste, build, dig, and wonder. They come away with dirty fingernails and dirty knees, often with pockets full of carrots, and a newfound connection to their own nourishment and the food system.

During the fall of 2009, Emily, Andy, and Dylan each partnered with a different fourth and fifth grade classroom on a "land and community" project. Their farms became extensions of the school building, and students spent time touring farms, harvesting, tasting, working, and asking any and every question they could think of. The unit combined science curriculum that included soil and geology, social studies curriculum focused on their own community, and a variety of extensions into literacy and math.

While the goal of the project was not to increase students' exposure to fresh produce or to change their eating habits, fennel quickly replaced McDonald's as a student favorite as their learning deepened. When students have the opportunity to spend time on farms, to get to know farmers and to see, taste, touch, and smell food in its natural state, they are far more willing to try it raw or prepared at mealtime, ask for it at home, incorporate it into their own diet and be willing to try new and different foods when they are offered. And these students were just getting their feet wet. Having Emily, Isha, Andy and Dylan around so frequently is a huge bonus because the students already know and trust them, and know that each of them brings incredible food into our school on a regular basis.

The farmers' support did not end with trips to their farms, either. Each farmer spent time with students in classrooms being interviewed about their careers and their role in the community. Dylan even brought a map of Intervale soil and explained to Mrs. Gordon's students how farmers decide what to grow where based on what the soil tells them.

As the weather turned colder and work on the farms slowed down, students began to take what they had learned about nutrition and their local food system and put it into practice. They ventured out into the markets in their own neighborhood, a tight knit low and middle income community that is home to new Americans from all over the world, Vermont natives, and transplants from around the country drawn to Burlington's progressive political climate. Students visited a Himalayan grocery store, an Asian market, two African halal stores, three convenience stores and the local food cooperative. At each store they looked for fruit and vegetables—local, fresh and processed—and began to parse out just where local and fresh produce was available, how much it cost, and what they could do with it once they got it home. They also began reading the young adult version of *Omnivore's Dilemma* (Pollan, 2009) during their literacy lessons.

In one classroom discussion about the book, a savvy fifth grader named Paulina raised her hand. "It doesn't make sense," she said, "It has to cost more to make a bag of Doritos than a bag of spinach. They have to grow all the ingredients, harvest them, process them, make the packaging, package them, make the advertisements, and then drive them to stores. And it still costs only a dollar, and you can get them anywhere on North Street, but spinach that Dylan grew right here in our town costs $5, and all he had to do was wash it, put it in a bag, and carry it up the hill, and you can't even get it in our neighborhood. And spinach is healthy, and Doritos *so* aren't."

So they asked Dylan why, took his answers, and began to think about how things could change. They also kept eating the veggies he and other farmers brought into school, taking slightly larger portions from the salad bar and slightly smaller amounts of barbeque sauce, and when a couple of heads of celeriac turned up one day, there were many eager tasters.

They also shared what they had learned with their community by hosting a community dinner and writing a book of poems, essays, prose and recipes titled *The Harvest Collection*. With special farmer guests, students prepared and served a meal to over 150 fellow students, family members, staff, and guests. Their meal, all homemade, included vegetable soup, roasted vegetables, salad, pasta, and carrot cake. As they served each guest, students were eager to share the ingredients and their origins with diners and why what they had on their plates were sound nutritional, economic, and ecological decisions. "Dylan grew that, but we harvested it," said one student, as a parent slid a small square of carrot cake onto her plane. "You can grow cake?" she asked. "Oh, no," Abdi replied, "he grew the carrots, and we harvested them. Then we made them into cake. You like it, don't you?" he said with a grin.

The Big Ideas for Success in Farm-to-School

All these stories are only the beginning of the incredible transition underway in Burlington. The success of the Burlington School Food Project is well known in the Farm-to-School world and is looked to as a model of what is possible. Harder than describing its many components is identifying the "big ideas" that reach across and beyond the action plans to the elements that have created the environment for such success. We present here, though, just some of the key components of a successful Farm-to-School program.

Partnerships and collaboration have been essential since the beginning of the program and continue to be. Strong relationships between city government, local farmers, non-profits, the school district, food businesses, local media, and citizens have assured that many ideas and many interests were being considered in all aspects of the project design and implementation. These relations have created a strong support network based on mutual respect and shared expectations. The current partners are always looking for new collaborators who might bring a different element to the work. "Making friends" with unlikely partners brings depth to the collaboration, especially remembering to take time to educate each other about your objectives, skills, and potential contribution.

Communication cannot be emphasized enough, and this means communication on all levels. Keeping all the partners engaged and up to date on project components as they move forward was initially achieved through monthly meetings and continues through regular meetings and constant emails. Sharing with everyone the story of the work that is under way, the work that has been accomplished, and the work that the project is aiming to achieve is important. Connecting with media resources, locally and nationally, keeps everyone engaged and can even bring in financial support. Probably most important is advertising any change made to the school menu. School menus and school newsletters have been the best place to highlight all the new foods being offered.

Focusing on group process and not the project's end product has created the environment and conditions for the project partners to tackle a host of issues. As the old saying goes, "Give a person a fish, you feed them for a day; teach them to fish, you feed them for a lifetime"; that could not be more true in Burlington. Instead of just deciding that we wanted to get, say, Vermont cheese or Vermont vegetables in all Burlington schools, we decided to focus instead on building a strong supportive team that could problem solve and grow together depending on what opportunities seemed most easily accomplished and the best use of our time. This brings us to the parallel element that our project partners use as an ongoing mantra, *staying nimble* and looking for the opportunities to present themselves. It is not about taking the curve balls when they come, but looking for them.

An element often forgotten in Farm-to-School programs that has been a fixed component in Burlington is constantly looking for opportunities for *student involvement*. This sometimes takes the form of educational activities, like classroom cooking and farm visits, but can also be involved in making changes with authentic experiences in the cafeteria. Students help in the kitchen, experimenting with food service and with new recipes. Students serve these new foods as taste tests for everyone to try. Students survey each other for honest feedback about new food items, and their feedback is used to alter recipes. Students have become a powerful voice of change, feeling empowered that what they think and say matters.

This feeling of empowerment reaches beyond the students, for *food service staff are empowered*, as well. A population within the workforce that historically did not feel they had much to contribute or were empowered to contribute are coming with ideas, problem-solving, and working toward a shared goal with the greater city community.

Most importantly, all project partners are *empowered to fail*. Identifying that our most important lessons learned were from our failures has freed all the partners to take risks and be confident that even if something doesn't work, we will learn a valuable lesson for the next time we try something. This freedom has allowed us to stay optimistic and taught us to be patient, as we know we are working on this in the long-term, not trying to solve everything today.

Looking back, if someone would have asked me what I thought we would have been able to accomplish between 2003 and 2010, I don't think I could have imagined our success or the far-reaching impacts they seem to have had. Moving forward, I feel the most critical thing to remember is that we don't know what we don't know. To me, this means we cannot look at a situation or suggestion and simply discount it with a "No, that will never work" or a "Our system doesn't support that" mentality. We are responsible for feeding hungry children every day, and we need to work to create change and to preserve and strengthen our food system. That makes each day an important one in the Farm-to-School process.

Reference

Pollan, M. (2009). *The omnivore's dilemma: The secrets behind what you eat* (Young readers ed.; adapted by R. Chevat). New York: Dial Books.

•CHAPTER NINE•

What's That Nonhuman Doing on Your Lunch Tray? Disciplinary Spaces, School Cafeterias, and Possibilities of Resistance

Abraham DeLeon

In a provocative introduction to Deleuze and Guatarri's (1983) *Anti-Oedipus: Capitalism and Schizophrenia*, the seemingly politically elusive Michel Foucault (1983) laid out a plan for political action in the context of that work. Calling the text a "book of ethics" (p. xiii), Foucault goes on to claim that it is our obsession with fascism as an ideology (not solely the German incantation during WWII) that is a challenge to new political movements: fascism as a "right-wing," authoritarian, intolerant, colonial, and exclusionary ideology (Mirasola, Sibley, Boca & Duckitt, 2007). Foucault (1983) provocatively asks, "how does one keep from being fascist, even (especially) when one believes oneself to be a revolutionary militant? How do we rid our speech and our acts, our hearts and our pleasures, of fascism?" (p. xiii). What Foucault points to is the pleasure of a fascist life, the lure of the Truths that a fascist life promises us. Fascism invades everything, from decisions made by the state to how larger discursive realities are performed by us every day. But Foucault's call for challenging the lure of fascism also evokes the ways this ideology is enacted on the bodies of nonhuman animals.

Human ideas about nonhuman animals arise in a variety of representations ranging from popular culture to Western forms of science and rationality (DeLeon, 2010a). Science, rigid classification systems, and positivism's role in the construction of narrow forms of truth arises from a fascist rationality that classifies (Wolfe, 2003), harvests (Nibert, 2002), gazes on (Chris, 2006), and

Others (DeLeon, 2010a) nonhuman animals for a bio-political control over life itself (Shukin, 2009). For educators, the problem of rethinking nonhuman animals in schooling emerges through spatial realities, the curriculum, and the representations found in the curriculum and textbooks. For example, the common scientific practice of dissection allows the expendable nonhuman body to be gazed upon and placed under the lens of Western forms of science and rationality. Although science would seem to be the domain of the nonhuman animal, the lack of representations in the social studies curriculum, for example, also shape our understanding of the role nonhuman animals should and could play in our society. Nonhuman animals also exist through various representations, often dressed as school mascots, embodying the discourses, ideologies, and subjectivities of human beings rather than some innate "animal" characteristic. Whatever the context, nonhumans become "objects," their bodies and representations existing to secure the hegemony of how our practices of domination are (re)produced. As these examples demonstrate, oppressive practices concerning nonhuman animals transcend official curricular boundaries, existing outside of the confines of the classroom and epistemological spaces that schools contain. The problem also exists in spaces like the cafeteria where nonhuman animals are unquestioningly served on the lunch trays of schoolchildren across the United States.

This issue is not new for education. There has been a recent surge in examining environmental concerns as well as nonhuman animals and their liberation, focusing on theory, praxis, and eco-justice in a variety of educational contexts (Andrzejewski, Pederson & Wicklund, 2009; Best, 2009; DeLeon, 2010a; Kahn, 2008, 2010; Kahn & Humes, 2009; Martusewicz & Schnakenberg, 2010; Pederson, 2009). However, this particular edited collection fills an important void in analyzing the roles of school food politics, especially the space of the cafeteria and its function in sustaining an oppressive relationship with nonhuman animals, a key component of a fascist life. Food is a fundamental feature of human culture, a "mixture of the organic and the inorganic, the material and the symbolic, the social and the natural" (Murdoch, 2006, p. 160). Thus, the ways in which we procure food for our society is a rich site of analysis and could be a point for opening up larger discussions about the nature of oppression within the United States.

This chapter builds a critique that focuses on the space of the cafeteria and its part in encouraging and sustaining meat consumption, especially because it is where we feed students nonhuman animals—and beliefs about humans' relationship to them. I give a short review of the literature on space, which places the cafeteria at the nexus of power and knowledge. Although it is recognized that activist work needs to challenge the lack of vegetarian/vegan options in cafeterias, there also needs to be more of a sustained attack against the various *ideologies* that support relationships of domination in the first place. Dismantling ideologies (especially anthropomorphism and the various discourses of speciesism) that shape institutional realities has to be taken as one of the ways in which a rigorous critique can be built for theory *and* praxis that resists the lure that fascism promises us. This chapter thus ends with ways to rethink the ideas of collective social action that scholars, teachers, administrators, and students might pursue to challenge hegemonic spaces like cafeterias.

Disciplinary Spaces, Anthropomorphism, Speciesism, and the Eating of Meat

Space is a fundamental feature of a socially constructed reality, a manifestation of how power, knowledge, and discourse create what is considered "real" or "true." Space informs not only reality, but also (re)produces relationships of power and dominant ways of thinking. "Recent works in the fields of literary and cultural studies, sociology, political science, anthropology, history, and art history have become increasingly spatial in their orientation" (Warf & Arias, 2009, p. 1). Space and ideology cannot be separated, and the ways in which the dominant social order has developed is intimately involved with the ways in which human beings have spatially ordered their lives. This means that space is central for not only the reproduction of capitalism, but also how supposedly neutral aspects of our everyday lives are enveloped in reproducing dominant ways of thinking about the world (Lefebvre, 1991).

For example, critical scholarship has located neoliberalism as an ideological and economic system that has had untold consequences for nature and our society (Shukin, 2009). Despite its economic roots, neoliberalism has also influenced diverse areas like incarceration (Brewer & Heitzeg, 2008), architecture (Harvey, 2009), public and private space (Mensch, n.d.), and education (Hursh, 2008). All of these are implicated within a complex web of *spatial* realities, taking place in school buildings, shopping centers, prisons, slaughterhouses, and public and private venues (Harvey, 2000). In this way, social problems have a spatial component, echoing the need to analyze space critically (Morgan, 2000). "A critical pedagogy of space would involve analyzing examples...to help students recognize the ways in which space is used to dominate and oppress some individuals and groups" (Morgan, 2000, p. 283). The structure of schools, the cafeteria, and the classroom all contain a spatial element, and this cannot be separated from the various technologies and discourses created that "produce" certain subjects, identities, and positionalities (Elden & Campton, 2007).

This organization of space occurs in subtle ways and is a fundamental feature of our contemporary society. Foucault was concerned with how power and space were manifested, specifically analyzing penal systems and the work of Bentham's architectural design of the *panopticon*. Although Bentham's architectural model for the prison appeared as the "eye of power," this also occurred outside of incarceration (Foucault, 1980a). Many aspects of social life are dedicated to the developing ways of closely monitoring and controlling bodies through surveillance technologies. During the eighteenth century in Europe, for example, architecture began to have a political end. "One begins to see a form of political literature that addresses what the order of a society should be, what a city should be, given the requirements of the maintenance of order" (Foucault, 1984, p. 239). Architecture became implicated with how the modern state would come to structure appropriate public and private space, ordering our bodies and daily existence.

Combined with the eighteenth and nineteenth century obsession with classifying and ordering the natural world, nature became a new obstacle for human development, and nonhuman animals would come to be feared, gazed upon, collected, dissected, caged, and domesticated (Serpell, 1996). This obsession with order (for humans and nonhumans alike) was at the forefront of the development of a disciplinary society. "Ordering is not just simply something we do, as when we make lists; more significantly, *it is something we are in*" (Hetherington, 1997, p. 35; emphasis added). Although the meaning of "order" is historically specific, it remains a pervasive aspect of how the West has organized itself under the influence of the Enlightenment (Foucault, 1972). Space, under specific ideological, historical and economic conditions, is vested in maintaining a strict social order, which controls movement, creates "appropriate" spaces for public and private activity, and structures reality insofar as space is "fixed" by the state to appear permanent and unchangeable. "Space is fundamental in any form of communal life; space is fundamental in any exercise of power" (Foucault, 1984, p. 252).

Schools are a vital component for the "spatial turn" experienced in the humanities and social sciences (Gulson & Symes, 2010). Schools are mired in a web of social and spatial practices that sustain commonly held and status quo approaches to time, bodies, labor, knowledge, production, and imagination (Morgan, 2000). In an era of high-stakes testing and accountability, the "business-as-usual" operation of schools now fits rigid and narrow requirements. With the support of various "accountability" programs to measure teacher performance, the gaze of the state and the social body is firmly set on teachers, who have to shoulder the burden of high-stakes testing and its classificatory mechanisms.

The pressures of standardized testing have fixed the goals of public schooling to a very narrow conception, preparing students for a specific logic of accountability and standardized normality. Discourses of accountability structure the spaces of classrooms, the pedagogy, and the curriculum that is supposed to meet this end product of "highly qualified" schools and bodies. These discourses also fix how other spaces are organized in school buildings. This means as critical scholars, we have to also examine the spaces outside the classroom—such as the cafeteria—and how they reproduce disciplinary and oppressive social practices in schools. For example, the library can be examined for its role in sustaining traditional and Eurocentric forms of knowledge production rather than critical forms of inquiry (Kumasi-Johnson, 2007), or, more to the purpose here, the cafeteria can become a site of critical inquiry into how oppression is both enacted and lived through what we decide to put on the plates of schoolchildren across the United States.

The cafeteria fills various roles in both the domination of nonhuman animals and the disciplinary practices that sustain this domination. A disciplinary society becomes manifest through the institutions created to manage and control a population (schools, prisons, hospitals, asylums), and it exists in the discursive regimes of a particular historical society (Foucault, 1972). "Discipline involves the organization of the subject in space through dividing practices, training, and standardization. It produces subjects by categorizing and naming them in a hierarchical order through a rationality of efficiency, productivity and 'normalization'" (Barker, 2008, p. 91). As a disciplinary space, the cafeteria exists to exert a certain power on the subject who passes through its doors. Discipline emerges through a variety of contexts as well. Whether that is manifested through the behavior management plans practiced on the bodies of students while they are eating in the cafeteria or the proliferation of fast-food corporations into high schools across the United States, students are exposed to a plethora of ideological and discursive realities concerning nonhuman animals and their apparent "suitability" for our endless consumption needs.

These types of disciplinary practices support discursive regimes and provide justification for a variety of social practices and institutions, ranging from the acceptance of disciplinary acts on our bodies through the internalization of rules and codes of conduct to the reality of corporations shaping everyday life and its possibilities. Discipline is a multi-faceted act that emerges through normalizing techniques such as statistical measures and classificatory systems that are rooted in Western and colonial ways of knowing (Barker, 2008). Discipline, then, exists in the ways that our knowledge and subjectivities are formed and regulated by the state (through biology, statistics, and classificatory systems of labels such as "mad" or "normal"). It exists as well in the actual practices that schools require their subjects to participate in and that they legitimize through daily practices and routines. The act of grabbing trays and utensils and standing in a kind of processing line to be fed seems to resemble how most people in the highly developed world experience their food reality. Although this is a lived experience that needs to be more fully analyzed from a phenomenological perspective, my concern has always been with the ways that ideology supports these types of practices and experiences in the first place. For example, there has been some focused activism on school lunches (e.g., The Center for Ecoliteracy, 2004), but little has been done towards advocating for vegan/vegetarian choices (People for the Ethical Treatment of Animals, PETA, n.d.; but see Weaver-Hightower's chapter in this volume). This gap demonstrates that the ideology and domination of nonhuman animals is such a facet of everyday life that it barely garners any attention, even within "progressive" reform efforts.

There is not only an almost complete invisibility of a vegetarian/vegan discourse within the larger conversations concerning cafeterias and student nutrition, but there is also an unspoken assumption that nonhuman animals are destined to end up as food for schoolchildren across the United States. A society that advocates and extols the beneficial aspects of meat consumption and hides the effects this meat-based economy has on the lives of nonhuman animals in factory farms and research facilities demonstrates the machinery needed to sustain this inherently unequal existence. This is even more apparent when we look comparatively at the practices of mass killings in factory farms and past historical injustices against human beings committed by those in positions of power (Spiegal, 1989). We seem to react in horror at the idea of human beings experiencing mass killings, but the invisibility of the death of untold numbers of nonhuman animals to the machines of capitalism still remains an acceptable practice (Davis, 2004).

The invisibility of nonhuman animals and their conditions of existence within larger discussions about health and the food we eat serves an important ideological function. Invisibility renders the subject meaningless, lost in a maze of discourses that lend justification to oppressive social practices and exclusionary politics; consider, for example, the invisibility of racism and the privilege that whiteness affords certain subjects (Sammel, 2009). Invisibility is a tactic utilized by corporations and large factory farms in hiding the devastating effects that meat consumption has on the environment and on the bodies of nonhuman animals, dressing up cows in smiles and living in happy family farms as the mainstream advertising of milk seems to suggest. The fast-food approach to what we eat within the context of the contemporary United States and other highly developed nations is a powerful reproductive mechanism, despite the detrimental health effects that these industries cause for our society (Schlosser, 2001). The cafeteria is a training ground for this invisibility, as meat comes packaged as patties, soups, tasty morsels and other aesthetic ways to hide the pain and suffering endured before it became that neat and tidy postmodern bio-package.

This type of ideological separation from nonhuman life outside of a consumer-based society is endemic to how modernism has also shaped humanity, especially the Fordist regime of production and proliferation. Raunig (2010) writes:

> Sociality in fordism implies the simultaneity of social subjection and solidarity as mutual dependency. Masses streaming into the Metro, uniformity and repetition, the punch-clock, the omnipresent *dispositif* of discipline and surveillance that constitutes the subjects as cogs in the fordist social machine. (pp. 10–11)

Unfortunately, this critical analysis omits how this cog can be applied to the arena of nonhuman animals. According to Foucault (1980b), a *dispositif* consists of:

> Discourses, institutions, architectural forms, regulatory decisions, laws, administrative measures, scientific statements, philosophical, moral and philanthropic propositions—in short, the said as much as the unsaid. Such are the elements of the apparatus. The apparatus itself is the system of relations that can be established between these elements. (p. 194)

If there exists a dispositif of disciplinary and surveillance practices for human societies and cultures, an equally disturbing dispositif exists for nonhuman animals, with various institutions, discourses, social practices, spaces, representations, beliefs, values, morals, and ideologies comprising our understanding of nonhuman animals. Schools—and within these larger spaces of domination the cafeteria and its own microcosm of power—exist in a web of discursive and social practices that supports vivisection, dissection, and consumption, all enacted on the bodies of nonhuman animals. Nonhuman bodies are rendered disposable and are constructed as profit-making mechanisms for fast food corporations and other consumer outlets. This dispositif also emerges through social practices as just discussed, but transcends practice to exist in the minds and ideas of the larger populace. Ideological justification has to exist to support this inequality, and for nonhuman animals, this is manifested through anthropomorphism and speciesism.

Anthropomorphism imagines that "animals think [and act] like humans," a form of what Daston and Mitman (2005) call a form of "self-centered narcissism: one that looks outward to the world and sees only one's own reflection mirrored therein" (p. 4). This has influenced not only representational practices (such as those that emerge in Disney films and other popular culture forms), but also practices that allow human beings to dominate nonhumans in research labs and lunchroom trays. "Humans project their own thoughts and feelings onto other animal species because they egotistically believe themselves to be the center of the universe" (Daston & Mitman, 2005, p. 4). For example, when we look at a dog, we will assign human qualities, maybe portraying his look as "sad." Or we may place nonhuman animals in our own roles as humans, such as in our occupations or familial roles. We position human beings at the top of a hierarchical pyramid representing the *value* of life, a troubling framework that allows a range of oppressive traditions to go unchecked and become normalized, producing "a range of subjects" (Barker, 2008, p. 92).

Along with naming anthropomorphism as an ideology rooted in domination, other scholars, like Gaard (2001), also point to speciesism, which can be defined as:

> a form of oppression that parallels and reinforces other forms of oppression. These multiple systems—racism, classism, sexism, speciesism—are not merely linked, mutually reinforcing systems of oppression: they are different faces of the same system. (p. 20)

Anthropomorphic and speciesist ideologies proliferate the space of the cafeteria, as students are given food that comes to their plates decontextualized. Humans come to see nonhuman animals as existing only as a means to an end for the sustenance of our bodies, with various scientific and nutritional discourses convincing us of the importance that meat and protein play in our diet as humans (Maley, forthcoming). Although these discourses may or may not be rooted in some form of truth, it does not excuse the oppressive means that we utilize to get to this end.

I also have argued elsewhere that the practices and ways in which human beings have *Othered* each other (as in Western European colonial and imperial projects in Africa and Southeast Asia, for example [Said, 1978]) strongly resemble the ways in which nonhuman animals are *Othered* physically, spatially, culturally, and ideologically (DeLeon, 2010a). This means that racism, classism, ableism, and other problematic "isms" are connected in a web of discursive and social realities and practices. The concept of *intersectionality* refers to the ways oppression operates effectively to not only suppress dissent and desire, but also to lend justification for unequal and oppressive social relationships that span race, class, gender, and other social markers. As the late French theorist Jean Baudrillard (1994) argued:

> Animals were only demoted to the status of inhumanity as reason and humanism progressed. A logic parallel to that of racism. An objective animal 'reign' has only existed since Man has existed. It would take too long to redo the genealogy of their respective statuses, but the abyss that separates them today, the one that permits us to send beasts, in our place, to respond to the terrifying universes of space and laboratories, the one that permits the liquidation of species even as they are archived as specimens in the African reserves or in the hell of zoos—since there is no more room for them in our culture than there is for the dead. (p. 133)

As the passage demonstrates, Baudrillard saw parallels to the treatment of marginalized humans with nonhumans, making the integral link between speciesism and other forms of oppressive realities. Baudrillard's critique highlights how intersectionality lies at the heart of how anthropomorphic and speciesist ideologies guide our exploitation and understanding of, and interactions with, nonhuman animals. But he also pushes us to recognize that Western forms of rationality, science, and humanism have created these rigid categories that allow the supremacy of human animals to remain unchecked. In general, science, psychology, history, and other academic disciplines have often reproduced the category of human as fixed and stable. But although humanism has tried to establish and solidify that category, there have been historical examples that have demonstrated that even "human" becomes dubious with the existence of pygmies and feral children (Nash, 2003). As Wolfe (2010) has argued, "persons aren't persons, in the sense of the definition of 'person' that humanism 'gives to itself'" (p. 119). Humanism positions us at the top of a hierarchical ladder of being and worth. This means that anything that does not match with our sense of experience, way(s) of knowing, or any other experiential marker that makes us "human" will be devalued at the expense of those "things" (plants or nonhuman animals, for example) that experience reality differently.

Despite the existence of these counter-narratives to how nonhuman animals are envisioned and exploited by human beings, the social practice of oppression exists firmly in the lives of everyday people. Especially troubling is the roles that these larger discursive and ideological realities play in the schooling of children, filtering to diverse realms such as the curriculum or the focus of this paper: the spaces where nonhuman animals are consumed, the cafeteria. But, establishing this critique of the practice of consumption in the cafeteria and the deconstruction of the ideologies that support these practices is not enough. The remainder of this chapter will be dedicated to returning to the question of fascism that Foucault offered and what can be done to resist the lure that it promises us as a collective social body.

Challenging the Lure of Fascism: Rethinking Spaces of Domination through Anarchist Theory

The ways nonhuman animals have been imagined and exploited within the West is a testament to how contemporary forms of fascism operate outside of the human milieu. From a series of ideologies that support domination to the actual practices of housing and slaughtering nonhumans for consumption, humans have tried to assert dominance over the nonhuman world. Tied to this are the spaces in which this domination occurs, like the factory farm, supermarkets, fast food, and public institutions. Space is a central feature to these practices, and the cafeteria is enveloped in these larger struggles for domination, serving up chickens, cows, and pigs to hungry students. However, nonhuman animals arrive on children's trays decontextualized, a non-entity that cannot reason, object, or give an alternative to the process through which it ended up on a lunchroom plate. The silence we accord nonhuman animals is deafening, and it remains a hidden and overlooked social process that occurs in schools. This authoritarian practice in which human beings routinely kill and abuse nonhuman animals is a great example of the lure of fascism that Foucault suggests seduces us into conformity. Because of this reality, we have to uncover new strategies of resistance, exposing the inherent contradictions of espousing to be a "free" society that lets these mass slaughters occur barely noticed.

Luckily, this silencing has been challenged in academic and activist circles (Best & Nocella II, 2004), a testament to the possibility that no matter how terrifying a system of power may be, resistance occurs (Foucault, 1983). Social change is a complex endeavor. Given the highly diffused nature of how reality is constructed, defying this contemporary arrangement becomes a challenge. In previous work, I have outlined ways in which we can rethink resistance through new subjectivities, exploring "self" through narrative inquiry, and building social critique through a rigorous deconstruction of ideologies, claims of truth, and dominant conceptions of the world (DeLeon, 2010b, 2010c, 2010d). In particular, I have highlighted anarchist theory as a possibility through which to build a diffused, nomadic, and *wild* resistance that subverts through direct action and sabotage (DeLeon, 2010d). In this tradition I have tried to establish ways in which activists and scholars can engage with anarchist principles of resistance while being located within a hierarchical and coercive institution like a public school or university.

Unfortunately, anarchism is routinely characterized as chaotic, unrealistic, and individual and seen as a general nuisance to law enforcement (Borum & Tilby, 2004). However, anarchists are concerned with building organic communities, finding new autonomous modes of being, rethinking relationships of power through a lens of direct action, and directly challenging oppressive state structures. For academics, anarchism also serves as an excellent way to critique hierarchies, ideologies of domination, and structures of power (Amster, DeLeon, Fernandez, Nocella II, & Shannon, 2009). Anarchists claim that it is not enough to see the space of cafeterias as oppressive or rooted solely in ideologies of domination; we must also think about ways in which subjects can resist how cafeteria space is organized and the various functions it serves. Space is an integral facet of everyday life in which these practices occur. If we problematize how this is constructed and (re)produced, we will be able to move away from a politics grounded in "truth" to one situated along a continuum of multiple truths, a multiplicity of affects that shift how subjects are discursively constructed (Clough, 2007).

Colebrook (2008) writes:

> Abandon the politics of truth, or an ethics of knowledge where we strive to know what we ought to do, and recognize the discursive forces and relations that produce truth. Truth and selves are effects of the relations of power . . . and is not some external substance that thought may adequately represent. (p. 36)

As a body of theory and way to think about praxis, anarchism attends to relationships of power because it lies at the heart of a coercive and hierarchical society, giving it a postmodern sensibility (Call, 2002; Newman, 2010). Anarchist conceptions of building communities, space, and resistance outside of a coercive and hierarchical totality should resonate with people concerned with nonhuman animals and how they are conceptualized and exploited in school cafeterias because of the hierarchies created in Western forms of science that classify and catalogue nonhuman animals and justify exploitation. This makes anarchist theory a nice complement for those wanting ideas about how to address practices concerning nonhuman animals.

Anarchism forces us to think differently about social issues and building trust, relying on communities that subscribe to some sort of communal or social justice paradigm to solve social problems for themselves. This framework eschews looking towards hierarchical and coercive institutions like the state for permission. For teachers, cafeteria workers, or nutritionists dedicated to resisting the status quo, this becomes a powerful framework in which to think about building collectives centered upon direct action and ideological subversion. Although Foucault and Deleuze were scant about specific political allegiances, their work is imbued with an anarchist sensibility, especially their ideas on power, knowledge, space, and bodies (Call, 2002). Their work forces us to de-center common assumptions found in the West and claims of truth embedded in socially sanctioned discourses. We can immediately see the educational implications.

If they so wish, teachers can conduct forms of ideological sabotage (which comes with risk as do any acts of resistance), exposing students to counter-narratives of how food is put on their plate, giving as much information as the teacher deems appropriate to facilitate discussions. PETA and other organizations have kits and books with which this can be done in ways that are productive for all involved. Student research into the issue also would facilitate this process, and through the creative act of writing (or other forms of media), students can transcend barriers and *become different*, subjects that are opened to new discourses of possibility that rethink old parameters and frameworks (Semetsky, 2006). This transformation can be facilitated through creative writing assignments that allow students to explore their own subjectivities, completing critical research assignments that uncover relationships of power or other ways of knowing, and forming activity groups that screen critical films, read challenging fiction, and explore the boundaries of their identities.

A transformation of self through ideological critique can change the relationship that we have with others, strengthening our own sense of community. As Ambrosio (2008) claims in the context of Foucault's work on the self:

> we transform ourselves by identifying the historical contingencies that have made us what we are, and by continually testing the cultural limits of what we can tolerate knowing, doing, and being. We cannot transform ourselves through a simple act of knowing, through critical reason or reflection alone, but only by risking who we are, by voluntarily seeking out and testing ourselves in situations that illuminate the contours of subjectivity, that destabilize our certainties. (p. 255)

Ambrosio is pushing us to recognize that along with challenging dominant power structures in whatever form that may take, we must test the limits of the self. Foucault (1983) pushes us to form politics of thought, action, and desire through a discourse of political action emerging from "proliferation, juxtaposition, and disjunction" (p. xiii). I discussed above how ideological critique is one facet, but exploring and actually *doing* other options with students must also occur.

For example, the current explosion of urban farming in cities like Detroit demonstrates that alternatives exist to mass-grown and processed food, putting our sustenance back into the hands of the community. This organic cohesion of life demonstrates intersectionality, connecting the self to nature through practice (gardening). In some schools, starting a garden could become a viable option, involving students in the process. These types of projects can expose students to different ways of thinking about how we procure food, transforming self in the process of exploring viable alternatives. By coming into contact with different perspectives and positionalities in regards to the food process, students can rethink relationships between human and nonhuman animals. This occurs through projects like gardening because it places students directly in the realm of food production, making them aware of the challenges of feeding our own communities. This type of experiential learning can demystify how food comes to our plates and eventually their lunchroom trays. Martusewicz and Schnakenberg (2010) write:

> Further, when students are encouraged to identify the aspects of their cultural and environmental commons that lead to mutual well-being and a smaller ecological footprint, they are offered the opportunity to understand these practices and relations as important assets needed to address current social and ecological injustices. (p. 29)

This type of rethinking and re-imagining challenges the diffused nature of reality and knowledge. Nonhuman animals are fixed within this social world, as Western forms of science and rationality have constructed them in very specific ways, such as automatons, unfeeling brutes, wild, savage, dangerous, feral, and troublesome. This does not mean that these representations are set for all time, but it is a reality that must be confronted when thinking about building forms of resistance against these pervasive representations.

Although humans do not yet have a language in which to speak to nonhumans that would legitimize their existence on their own terms, humans must simultaneously also think about our own transformation by engaging the "question" of the animal. In this way, transforming our own discourses and subjectivity could lead to a wider exploration of possibilities of existence that stand outside our current social, economic, and political parameters of domination. As this chapter has demonstrated, this domination is rooted in Western forms of science and rationality. I specifically locate this in the West merely because I reside in the Western Hemisphere, and Western forms of thinking have significantly influenced how states like the United States have developed. The West has had an ambivalent or openly hostile relationship with nonhumans, one that has influenced the ways in which we view the process of our food sources, shaping relationships for children and their interactions with nonhuman animals as their source of food.

This last point cannot be overlooked, as students are given few choices outside factory-farmed food, mirroring very specifically the ways that curriculum is also developed and delivered to students. Social practices in schools must include rethinking the possibilities that a public education can offer, but also the learner's role in (re)producing and resisting the standards and norms prescribed by a disciplinary society. That society has to move towards "difference over uniformity, flows over unities, [and] mobile arrangements over systems" (Foucault, 1983, p. xiii). In other words, we need to rethink the political possibilities available to us as educators during the era of late capitalism through a politics rooted in ecological understanding, connecting diverse forms of knowledge, practices, discourses, and institutions (Weaver-Hightower, 2008). The tendency of education to focus strictly on the "human" condition becomes problematic as spaces like school cafeterias fulfill a role in reproducing much larger ideologies and discourses that support the mass killing and extermination of nonhuman animals that have far-reaching effects outside of the school and cafeteria walls.

Towards Moments of Resistance: Rethinking the Cafeteria Space

Space cannot be underestimated when thinking how we reproduce relationships of power through a variety of problematic social practices. Returning to Foucault's call to challenge the emergence and lure of fascism in any form it takes, critical educators interested in building a society based upon principles of social justice are compelled to look outside the human milieu to explore how oppression is enacted on those who seem to occupy the lowest rungs of our collective social world. Nonhuman animals that are deemed "food" (through a variety of oppressive practices and ideologies of domination), are exploited, tortured, and killed, and this vicious cycle ends on the trays of millions of schoolchildren across the United States. Through disciplinary practices and the arrangement of space, the cafeteria becomes a conduit for these larger ideologies that support vivisection, dissection, experiments, and factory farming. Students, teachers, nutritionists, and administrators are enveloped in these social practices, often without any knowledge of what occurs to nonhuman animals that end up destined to be our food. This edited collection, however, forces us to think about the politics of food and the implicit ways in which we participate in this reproduction.

Although the school is coercive at all of its various levels (from the curriculum to the disciplinary practices that keep students "behaving"), resistance is possible. In whatever ways resistance emerges, it should be envisioned by the local community within which the school is situated, building an organic, localized resistance against the plight of nonhuman animals in our society, but also revising ways of thinking about nonhuman animals. However, praxis must be explored within a critical theoretical framework that builds critique along with ways that students can affect change. By combining social theory and praxis, we can think of resistance like its own ecological system. Thinking about educational policy through an ecological framework crosses institutional, lived, and discursive realities, making it difficult to pinpoint an origin or center. By thinking about social movements this way, we can think about how the nonhuman animal would be one of the "actors" of this ecological system, an important link in how this ecology operates (Weaver-Hightower, 2008). A vital component of this ecology is not only the science laboratory down the hall, but also the place where we nourish the bodies of our students. Suffering matters, and although many may be oblivious to how the meat arrives on their tray, we cannot hope for any larger social change to occur within the "human" milieu when such widespread suffering happens at the hands of our own social practices. By examining these, critiquing them, and conducting direct actions, we aim for a better future that recognizes that beings must be able to chart their own destinies, despite our inability (or sometimes unwillingness) to understand how this phenomenon occurs.

References

Ambrosio, J. (2008). Writing the self: Ethical self-formation and the undefined work of freedom. *Educational Theory, 58*(3), 251–267.

Amster, R., DeLeon, A., Fernandez, L., Nocella II, A. & Shannon, D. (Eds.). (2009). *Contemporary anarchist studies: An anthology of anarchy in the academy*. London, UK: Routledge.

Andrzejewski, J., Pederson, H. & Wicklund, F. (2009). In J. Andrzejewski, M. Baltdodano & L. Symcox (Eds.), *Social justice, peace and environmental education: Transformative standards* (pp. 136–154). New York: Routledge.

Barker, C. (2008). *Cultural studies: Theory & practice*. Thousand Oaks, CA: Sage.

Baudrillard, J. (1994). *Simulacra and simulation*. (S. F. Glaser, Trans.) Ann Arbor: University of Michigan Press.

Best, S. (2009). The rise of critical animal studies: Putting theory into action and animal liberation into higher education. *Journal for Critical Animal Studies, VII*(1), 9–53.

Best, S. & Nocella II, A. (Eds.). (2004). *Terrorists or freedom fighters? Reflections on the liberation of animals*. New York, NY: Lantern.

Borum, R. & Tilby, C. (2004). Anarchist direct actions: A challenge for law enforcement. *Studies in Conflict and Terrorism, 28*(3): 201–223.

Brewer, R., & Heitzeg, N. (2008). The racialization of crime and punishment: Criminal justice, color-blind racism, and the political economy of the prison industrial complex. *American Behavioral Scientist, 51*(5), 625–644.

Call, L. (2002). *Postmodern anarchism*. Lanham, MD: Lexington Books.

Center for Ecoliteracy. (2004). *Road map: Rethinking school lunch guide*. Berkeley, CA: Learning in the Real World.

Chris, C. (2006). *Watching wildlife*. Minneapolis: University of Minnesota Press.

Clough, P. (2007). Introduction. In P. Clough (Ed.), *The affective turn: Theorizing the social* (pp. 1–33). Durham, NC: Duke University Press.

Colebrook, C. (2008). Leading out, leading on: The soul of education. In I. Semetsky (Ed.), *Nomadic education: Variation on a theme by Deleuze & Guattari* (pp. 35–42). Rotterdam, NL: Sense Publishers.

Daston, L., & Mitman, G. (2005). Introduction. In L. Daston & G. Mitman (Eds.), *Thinking with animals: New perspectives on anthropomorphism* (pp. 1–14). New York, NY: Columbia University Press.

Davis, K. (2004). A tale of two holocausts. *Animal Liberation Philosophy and Policy Journal, 2*(2), 1–20.

DeLeon, A. (2010a). The lure of *The Animal*: The theoretical question of the nonhuman animal. *Critical Education, 1*(2). Retrieved August 1, 2010, from http://m1.cust.educ.ubc.ca/journal/index.php/criticaled/article/view/71/123

DeLeon, A. (2010b). Reporting from the realm of the absurd: Rethinking space in a neoliberal world. *Theory in Action, 3*(4).

DeLeon, A. (2010c). How do I tell a story that has not been told: Anarchism, autoethnography and the middle ground. *Equity & Excellence in Education, 43*(4), 398–413.

DeLeon, A. (2010d). Anarchism, sabotage and the spirit of revolt: Injecting the social studies with anarchist potentialities. In A. DeLeon & E. Wayne Ross (Eds.), *Critical theories, radical pedagogies and social education: New perspectives for social studies education* (pp. 1–12). Rotterdam, Netherlands: Sense Publishers.

Deleuze, G. & Guattari, F. (1984). *Anti-Oedipus: Capitalism and schizophrenia.* (R. Hurley, M. Seem & H. Lane, Trans.). Minneapolis: University of Minnesota Press.

Elden, S. & Crampton, J. (2007). Space, knowledge and power: Foucault and geography. In S. Elden & J. Crampton (Eds.), *Space, knowledge and power: Foucault and geography* (pp. 1–18). Burlington, VT: Ashgate Publishing.

Foucault, M. (1972). *The archaeology of knowledge & the discourse on language.* (A. M. Sheridan Smith, Trans.). New York, NY: Pantheon.

Foucault, M. (1980a) The confession of the flesh. In C. Gordon (Ed.), *Power/knowledge. Selected interviews and other writings 1972–1977* (pp. 194–228). New York, NY: Pantheon.

Foucault, M. (1980b). The eye of power. In C. Gordon (Ed.), *Power/knowledge. Selected interviews and other writings 1972–1977* (pp. 146–165). New York, NY: Pantheon.

Foucault, M. (1983). Preface. In G. Deleuze & F. Guattari, *Anti-oedipus: Capitalism and Schizophrenia* (R. Hurley, M. Seem & H. Lane, Trans.). Minneapolis: University of Minnesota Press.

Foucault, M. (1984). Space, knowledge, power. In P. Rabinow (Ed.), *The Foucault reader* (pp. 239–256). New York, NY: Pantheon Books.

Gaard, G. (2001). Ecofeminism on the wing: Perspectives on human-animal relations. *Women and Environments International Magazine, 52*(53), 19–22.

Gulson, K. & Symes, C. (2010). Knowing one's place: Educational theory, policy, and the spatial turn. In K. Gulson & C. Symes (Eds.), *Spatial theories of education: Policy and geography matters* (pp. 1–16). New York, NY: Routledge.

Harvey, D. (2000). *Spaces of hope*. Berkeley: University of California Press.

Harvey, D. (2009). *Cosmopolitanism and the geographies of freedom*. New York, NY: Columbia University Press.

Hetherington, K. (1997). *The badlands of modernity: Heterotopia & social ordering*. London, UK: Routledge.

Hursh, D. (2008). *High-stakes testing and the decline of teaching and learning: The real crisis in education*. Lanham, MD: Rowman & Littlefield.

Kahn, R. (2008). Towards ecopedagogy: Weaving a broad-based pedagogy of liberation for animals, nature, and the oppressed peoples of the earth. In A. Darder, R. Torres, & M. Baltadano (Eds.), *The critical pedagogy reader* (2nd ed.; pp. 522–540). New York, NY: Routledge.

Kahn, R. (2010). *Critical pedagogy, ecoliteracy, and planetary crisis: The ecopedagogy movement*. New York, NY: Peter Lang.

Kahn, R. & Humes, B. (2009). Marching out from Ultima Thule: Critical counterstories of emancipatory educators working at the intersection of human rights, animal rights, and planetary sustainability. *Canadian Journal of Environmental Education, 14*, 158–178.

Kumasi-Johnson, K. (2007). Critical inquiry: Library media specialists as change agents. *School Library Media Activities Monthly, 23*(9), 42–45.

Lefebvre, H. (1991). *The production of space*. Oxford, UK: Blackwell.

Maley, C. (Forthcoming). Meet them at the plate: Reflections on the eating of animals and the role of education therein. *Critical Education*.

Martusewicz, R. & Schnakenberg, G. (2010). Ecojustice, community-based learning and social studies education. In A. DeLeon & E. W. Ross (Eds.), *Critical theories, radical pedagogies and social education: New perspectives for social studies education* (pp. 25–42). Rotterdam, NL: Sense Publishers.

Mensch, J. (n.d.). Public space. Retrieved September 16th, 2010, from http://www.europhi losophie.eu/mundus/IMG/doc/Public_Space-Mensch.doc

Mirasola, A., Sibley, C., Boca, S. & Duckitt, J. (2007). On the ideological consistency between right-wing authoritarianism and social dominance orientation. *Personality & Individual Differences, 43*(7), 1851–1862.

Morgan, J. (2000). Critical pedagogy: The spaces that make the difference. *Pedagogy, Culture and Society, 8*(3), 273–289.

Murdoch, J. (2006). *Post-structuralism geography: A guide to relational space*. Thousand Oaks, CA: Sage.

Nash, R. (2003). *Wild enlightenment: The borders of human identity in the eighteenth century*. Charlottesville: University of Virginia Press.

Newman, S. (2010). *The politics of postanarchism*. Edinburgh, UK: Edinburgh University Press.

Nibert, D. (2002). *Animal rights/human rights: Entanglements of oppression and liberation*. Lanham, MD: Rowman & Littlefield.

Pederson, H. (2009). *Animals in schools: Processes and strategies in human-animal education*. West Lafayette, IN: Purdue University Press.

People for the Ethical Treatment of Animals. (n.d.). Retrieved September 1st, 2010, from http://www.peta.org/action/default.aspx

Raunig, G. (2010). *A thousand machines: A concise philosophy of the machine as social movement.* Los Angeles, CA: Semiotext(e).

Said, E. (1978). *Orientalism.* New York: Vintage.

Sammel, A. (2009). Turning the focus from 'Other' to science education: Exploring the invisibility of whiteness. *Cultural Studies of Science Education,* 4(3), 649–656.

Schlosser, E. (2001). *Fast food nation: The dark side of the all American meal.* New York: Houghton Mifflin.

Semetsky, I. (2006). *Deleuze, education and becoming.* Rotterdam, Netherlands: Sense Publishers.

Serpell, J. (1996). *In the company of animals: A study of human-animal relationships.* Cambridge, UK: Cambridge University Press.

Shukin, N. (2009). *Animal capital: Rendering life in biopolitical times.* Minneapolis: University of Minnesota Press.

Spiegel, M. (1989). The dreaded comparison: Human and animal slavery. New York: Mirror Books.

Warf, B. & Arias, S. (2009). Introduction: The reinsertion of space into the social sciences and humanities. In B. Warf & S. Arias (Eds.), *The spatial turn: Interdisciplinary perspectives* (pp. 1–10). New York, NY: Routledge.

Weaver-Hightower, M. (2008). An ecology metaphor for educational policy analysis: A call to complexity. *Educational Researcher,* 37(3), 153–167.

Wolfe, C. (2003). *Animal rites: American culture, the discourse of species, and posthumanist theory.* Chicago, IL: University of Chicago Press.

Wolfe, C. (2010). *What is posthumanism?* Minneapolis: University of Minnesota Press

2

• CHAPTER TEN •

Coda
Healthier Horizons

Sarah A. Robert
Marcus B. Weaver-Hightower

"But Mommy, I don't need to eat breakfast; they give it to me in my classroom." My (Sarah's) four-year-old did not want to finish her oatmeal. I was afraid to hear such a statement. I was aware that the urban and poor school district she attended had begun to provide breakfast in the classroom through some impenetrable-acronymed program to "serve" the impoverished city's youngest inhabitants. I was very aware of the great need for public food assistance, but I was suspect too. "What do you get to eat for breakfast at school?," I asked. "Trix, cookies, milk, and juice." "Sit down and eat," was my response. I thought of the diabetic coma that such a breakfast could induce or, worse, how it could drive a classroom of pre-kindergartners to energetic highs. The weather was getting cooler now, too. Students would not be taken outside to play; no recess would be offered until March or April. A few weeks later, my nine-year-old brought home a letter introducing a new healthy eating curriculum including permission slips for fieldtrips to a local supermarket and a play about healthy choices. We were asked to sign off on our daughter's participation and "to promise" that we would review materials provided online either at home or our nearest public library so that we, her parents, would learn about healthy choices, too.

The junk food breakfast, minimal physical activity, and the healthy eating curriculum are found in the same school. The contradictory messages illustrate the paradoxes of current school feeding and nutritional education ecology in the United States and many other national contexts. The message is that we'll give you refined sugars, supposedly so that you can learn better, but we will give no avenue to burn them off. And that same processed food you "choose" to eat—my four-year-old could refuse it, presumably—will destroy your body and inhibit learning, as the healthy eating play revealed to my nine-year-old daughter. Both programs uniquely engage perverse socio-cultural and political notions that students have control and must make wise choices, while at the same time food politics involving multiple levels of governance shape what exactly those choices are. This book has juxtaposed school food and politics to not only illustrate school feeding ecologies but also to illustrate how people and groups challenge school feeding as it is organized and practiced at the beginning of the twenty-first century.

In this final chapter, we revisit the aims of the collection and the key terms of food politics and policy ecologies that unify our diverse cases. We also synthesize the "food for thought" that we hope the reader has garnered, specifically: that reforming school food needs to be a much broader and complex endeavor that can—and should—contemplate the ecology within which food is produced, distributed, consumed, and learned about; that it is crucial to contemplate the question "*who feeds whom, what, how, and for what purpose*" to understand and critique that ecology; and, last, that strategies for changing school food ecologies come from such analyses and from work on the ground in the ecology.

Restating Aims and Key Terms

School food is political. And school food involves politics. In fact, from a reading of the various chapters, school food involves politics as defined by various theorists. Paarlberg's (2010) assertion that politics are the struggles of organized actors to engage the state over food regulation applies. Chapters illustrated how groups from Australia to South Korea have pushed governments on the quality and distribution of food to students and even the purpose of school feeding in and of itself. Indeed Gee's (2005) notion of politics as the distribution of social goods is evident in this compilation's pages, too. Whether it is the collaborative work of the Burlington School Food Project or the UK-based School Food Trust, myriad groups are working to bring food and food knowledge to students, distributing a social good that can extend beyond a full belly. There also is Connell's (2009) assertion that politics involve struggles over limited material and symbolic goods. From Argentina to Tanzania to the United States, groups of parents, teachers, and community members are struggling against unequal distribution of school food and the social goods that come with nutrition: physical health, improved learning, human rights, dignity. Still, politics also can involve working outside of the direct constraints of oppressive state authority or hierarchical institutions to tackle the symbolic violence of everyday life (Bourdieu & Passeron, 1990). Cases from the United States also illustrate the possibilities for engaging issues of food, equity, and liberation from the ground up. As both the title of this collection and its contents suggest, school food politics are widespread and multiplex.

Contemplating strategies to change school food has required an expansive view of the context being scrutinized. Juxtaposing "school," "food," and "politics" has led authors in this collection to follow the many links in the political food chain, both so they could understand it and then, in many cases, transgress it. One crucial strategy has been to look to *history*. Many of the authors started their narratives from a historical perspective to critique the past and the present and to construct healthier horizons for school feeding. This important groundwork avoids historical amnesia (see Phillips & Roberts, this volume), setting the stage for ongoing and future engagements with the topic, whether in academic research or on-the-ground activism. Tracing out program goals, origins, traditions, funding, and the groups concerned with controlling school feeding reveals the problems and possibilities for change.

Another strategy employed by contributors for understanding and changing the way schoolchildren were fed was to seek a deep understanding of the complex participation of *actors*. Multiple groups of actors affect policy processes, at times from different levels within an ecology. Community-based groups and parents, for example, struggle for voice over what is fed to their children. How parents and community groups engage the state brings in another group of actors working within a different level of the state apparatus. As important as ascertaining the effects actors have on school food, though, is pausing to consider the work researchers and activists do reflecting on research and advocacy practices within the ecology, work done on behalf of healthier horizons for students. Beyond analysis and even beyond answering the question of *who feeds whom, what, how, and for what purpose*, researchers and activists must answer another question, So what?! And with the response—whatever it might be—scholars and advocates seek out means to engage in politics to improve school feeding, to do the work of public intellectuals and community activists.

The strategy of examining *space* also proved crucial for understanding the complex terrain of food policy. School feeding involves multiple policies and practices emanating from multiple spaces. Though many programs are anchored by a national-level school feeding policy, the current context of decentralization means that decision making does not stop at the federal level but rather is diffuse, complicating the process of change while also opening up promising spaces in which to act. Yet transnational agreements and the non-

Euclidean "space" of globalization also influence how and what a nation and even a local community feeds its children. Researchers of school food politics must seek out and occupy these diverging spaces to capture the dynamic relationships and processes behind the food on a lunch tray. Fixing any school food ecology is both a local and a global project. This compilation is a call to researchers and activists alike to learn about various places and spaces, often to step outside of one's comfort zone to piece together a school food ecology that embraces and respects these interconnections.

All of these strategies lend themselves to envisioning school food policy as an ecosystem. Applying a policy ecology framework to understand and then transform school food is of course only one possible approach among many. Still, the authors contemplated and applied—to varying degrees of explicitness—Weaver-Hightower's (2008) framework to delve into or broaden the scope of their analysis, demonstrating the analytical power that such thinking can have for understanding an immensely complicated (some might say Byzantine) object like the global food system. Even so, much complexity remains to be uncovered.

What We Have Learned About Possibilities for Healthier Horizons

Broadening Policy Horizons

As the editors of this compilation, our vistas of school food ecologies have broadened significantly. We have learned from the various contributors to think about, critique, and just view food, food education, and the space in which it is served in schools in a new light. We also have been given pause to rethink the terrain we currently occupy. Policy analysis, for one, needs to be a much broader and complex endeavor. Policy analyses can—and should—contemplate the ecology within which food is produced, distributed, and consumed. School food policy is not an "objective" or a limited scope endeavor. Vast amounts of money are involved or should be, pulling in even the most unlikely actors to decision-making processes about school feeding. Decisions about what is served to whom are influenced somewhat by local communities. They are also influenced by those wielding control of the funding, and that control is shaped by politics that have for generations been characterized by inequality. So further than broadening the scope of school food policy analysis beyond who acts on feeding, analyses must acknowledge any school feeding as an *ideological* project. Who feeds students, what students eat, and whether they eat reveal ideological interventions. Eating in the cafeteria is not a benign, apolitical activity. Children are taught a curriculum for life in the cafeteria, canteen, or school patio.

One particular vista that has emerged in the previous pages merits further elaboration: school food's relation to rights-based discourses. While human rights discourses are not without their criticism, they resonate with the demands of so many of the actors involved in food struggles. Nutritious food "should" be considered a human right of all children connected to the right to an education. We know that children learn better on a full belly. We know children will attend school when fed. Why isn't a child's right to food a right of citizenship too? As a member of a society, one "should" have the right to live and thrive in that same society. Moving away from economic rationales for school feeding decisions should continue with a re-centering on the rights of children and their communities.

Asking Questions About the School Food Ecology

Asking who feeds whom, what, how, and for what purpose reveals a great deal about how children are cared for by a society. Sandler posed these questions as a means to enter into an ecology to critique it from a deep understanding of lived experiences enveloped in politics. At the beginning of the twenty-first century, those lived experiences are often framed discursively and materially by crisis politics. Crisis and opportunity, however, are sometimes considered synonymous. Pablo Pineau (2003), an Argentine educational researcher, proposes that in times of crisis educators contemplate and learn from others, to become or return to being learners. He suggests learning from community members not usually deemed knowledgeable by dominant groups, based on power matrices of gender, class, race, ethnicity, and nationality. Pineau, for example, suggests educators (really, any political actor) seek out the knowledge of women in poor communities. How do they manage to feed so many in moments of crisis? What work is entailed in doing so? In this example, we see that not only should an understanding of how to fix school feeding entail an understanding of the identity politics running through community feeding projects, but also we see a need to acknowledge the labor involved in school feeding. This school food work is gendered, raced, and classed. As chapters in this compilation have similarly shown, identity politics run through school food programming, and we learn much from these chapters and the diverse actors profiled in them that can be used to make school food healthier and fairer.

Reality checks are important, but so too is imagining alternatives to the current context of school feeding. In this regard we take inspiration from the powerful motto of the World Social Forum: "A New World Is Possible." Questioning the current school food ecology should be accompanied by a vision of what it ought to be. Knowing who feeds whom, what, how, and for what purpose—and knowing the sociocultural systems of inequality that frame the what, how, and why—should be a starting point, not the end goal. A new school food ecology is indeed possible. As the authors showed in this book, healthier horizons are not just imagined; they are struggled for, in some cases for decades. Alternatives to entropic, fragmented, oppressive, nonexistent, or just unhealthy school feeding programs exist in small and large communities from Australia to South Korea to Burlington, Vermont. Many communities, however, continue to struggle for the equitable distribution of resources to feed their children. Many different actors in a school food ecology must struggle together, but they must also engage the state—often multiple governing bodies—in order to engage in food politics (Paarlberg, 2010).

Strategies for Changing School Food Ecologies, Strategies for Educating

Schools, in their current realization, cannot (and perhaps should not) be all-encompassing state institutions tasked with resolving all social problems. Nevertheless, schools do a lot! What is important to note is that "many times it is not possible to repair what has been broken in other places" (Pineau, 2003, p. 117). In other words, schools are not in a position to fix problems caused not by themselves but by the failures and oversights of politicians, the economy, militaries, and civil societies. Pineau suggests, however, that it is possible to do what schools have been asked to do—or that they have taken on out of concern for educating—and do the tasks well.

We argue that feeding students nutritious food should not be considered an added burden or an extraneous philanthropic task. School feeding is not a short-term development project, or a fix in a current "crisis." It is, instead, an integral part of educating. And the objective should be for schools to educate all students. One approach for meeting that objective is to be aware of the symbiotic relation of food and education and then to craft policies that link the two. School food policy can help education and education policy can help food. As such, fixing the school food policy ecology should be the "educational project of nations" (Weaver-Hightower, this volume). A deep, critical look at school food policy tells us much about the education schools provide. It shows how little we genuinely want children to focus and learn when we do not feed them nutritious food to fortify them throughout the school day. It shows how much a society cares about children's long-term growth beyond the school day, across years. Healthier meals help schools do what they should be focused on: educating during the school year and cultivating lifelong learning. One of the food-related sayings woven into this book—"you are what you eat"—should be contemplated as part of any local or national educational project. If this saying is true, it has tremendous implications for how we both feed and educate. Who do we as a society want our children to be? Given that, what should we be feeding—nutritionally and academically—all of our children? Stopping to consider the cafeteria or a look into a classroom serving breakfast can help answer these questions.

School food politics and their implications for education should be considered in teacher education programs, continuing education programs, administrator training programs, and in school curricula at every level; however, that curriculum should not sit in contradiction with the food served students. We believe teachers know a lot. Yet, how much are teachers taught about their students and their communities, and how much are they taught about how to get to know them through everyday contact and research? How much are teachers taught about the structures of poverty, the constraints of school funding, and education politics? How many administrators encourage collaborations across disciplines or between the classroom and cafeteria? Knowing content and pedagogy are crucial for learning but so too is knowing the context in which one educates. We implore fellow educators to (continue to) value learning about education politics, foundations, and the world beyond the classroom as crucial to the education of teachers, administrators, and—by extension—the students both work to educate.

As deceptively simple as school food may appear, reforming school food ecologies involves complex politics. Despite the complexity involved in reform, healthier and more just horizons are possible for schoolchildren. Burlington's school food reforms (Davis et al., this volume), as just one example, demonstrate a compelling vision of healthier and more just school food, and they provide concrete ways to achieve it. As such programs show, communities must educate themselves about schools, food, and politics, and they must collaborate to chart a path to new horizons.

References

Bourdieu, P., & Passeron, J. C. (1990). *Reproduction in education, society, and culture* (R. Nice, Trans.). London, England: Sage.

Connell, R. W. (2009). *Gender.* London: Polity Press.

Gee, J. P. (2005). *An introduction to discourse analysis: Theory and method* (2nd ed.). New York, NY: Routledge.

Hodgson, D. (2001). *Once intrepid warriors: Gender, ethnicity, and the cultural politics of Maasai development.* Bloomington: Indiana University Press.

Paarlberg, R. (2010). *Food politics: What everyone needs to know.* Oxford, England: Oxford University Press.

Pineau, P. (2003). O escuela o crisis. Crónicas marcianas del imaginario docente actual. In I. Dussel & S. Finocchio (Eds.), *Enseñar hoy: Una introducción a la educación en tiempos de crisis*

(Teaching today: An introduction to education in times of crisis) (Primera ed., pp. 113–118). Buenos Aires, Argentina: Fondo de Cultura Económica.

Weaver-Hightower, M. B. (2008). An ecology metaphor for educational policy analysis: A call to complexity. *Educational Researcher, 37*(3), 153–167. doi: 10.3102/0013189X08318050

• APPENDIX •

School Food Resources
From Curriculum to Policy to Recipes

The purpose of this appendix is to offer readers a variety of resources from which they can learn more about school food politics. These include websites, curricula and lesson plans, and scholarly articles and books. We hope these will supplement the wonderful resources cited in the individual chapters in this collection.

Websites

Asterisks (*) mark websites that offer recipes; check marks (√) indicate websites that offer lesson plans or curricula.

Better School Food
http://www.betterschoolfood.org/

Better DC School Food*
http://betterdcschoolfood.blogspot.com/

Centers for Disease Control and Prevention (CDC) Healthy Youth Program
http://www.cdc.gov/HealthyYouth/

Center for Food and Justice, Occidental College, Urban and Environmental Policy Institute
http://departments.oxy.edu/uepi/cfj/lausd.htm

Center for Ecoliteracy √
http://www.ecoliteracy.org/

Chef Ann Cooper, the Renegade Lunch Lady *
http://www.chefann.com/

Chez Panisse Foundation √
https://www.chezpanissefoundation.org/school-lunch-reform

Edible Schoolyard* √
http://www.edibleschoolyard.org

Farm to School √
http://www.farmtoschool.org/

Feeding Minds, Fighting Hunger √
http://www.feedingminds.org/default.htm

Fed Up With Lunch
http://fedupwithschoollunch.blogspot.com/

Food First, Institute for Food and Development Policy
http://www.foodfirst.org

The Food Museum Online
http://www.foodmuseum.com/index.html

Food Politics blog by Marion Nestle
http://www.foodpolitics.com

The Food Project √
http://www.thefoodproject.org

The Food Studies Institute √
http://www.foodstudies.org

The Food Trust
http://www.thefoodtrust.org/index.php

Healthy Schools Campaign
http://www.healthyschoolscampaign.org/

Healthy School Lunches * √
http://www.healthyschoollunches.org/

Jamie Oliver's Food Revolution * √
http://www.jamieoliver.com/foundation/jamies-food-revolution/

Let's Move! Campaign* √
http://www.letsmove.gov/

Local Area Caterers Association (UK)
http://www.laca.co.uk/

The Lunchbox * √
http://www.thelunchbox.org

The Lunchtray
http://www.thelunchtray.com

National Coalition for Food-Safe Schools
http://www.foodsafeschools.org/

One Tray
http://onetray.org/

Organic Consumers Association
http://www.organicconsumers.org/

Rachel Ray's Yum-O! Non-profit Cooking Program*
http://www.yum-o.org/

Rudd Center for Food Policy and Obesity
http://www.yaleruddcenter.org/

School Food Focus
http://www.schoolfoodfocus.org/

School Food Trust (UK) * √
http://www.schoolfoodtrust.org.uk/index.asp

School Lunch Talk
http://www.schoolfoodpolicy.com/

School Nutrition Association
http://www.schoolnutrition.org/

The s'Cool Food Initiative
http://www.scoolfood.org/

The Slow Cook
http://www.theslowcook.com/

Slow Food USA's Time for Lunch Campaign
http://www.slowfoodusa.org/index.php/campaign/time_for_lunch/

USDA School Meals
http://www.fns.usda.gov/cnd/

World Food Program √
http://wfp.org

Curriculum and Lesson Plans

Center for Ecoliteracy. (2010). Rethinking school lunch guide. (2nd Ed.) Berkeley, CA: Center for Ecoliteracy. Retrieved from http://www.ecoliteracy.org/sites/default/files/uploads/ rethinking_school_lunch_guide.pdf

Chez Panisse Foundation. (2010). Evaluation of the school lunch initiative: Changing students' knowledge, attitudes, and behaviors in relation to food. Retrieved from https://www.chezpanissefoundation.org/ uploads/file/sli_exec%20sum_100921.pdf

Feeding a Hungry World: Focus on Rice in Asia and the Pacific. SPICE.stanford.edu

Food Standards Agency. (2009). *Food route: A journey through food.* Retrieved from http://www.food.gov.uk/multimedia/pdfs/foodrouteprimuser.pdf

Kansy, H. (2003). *Food connections in world history: Contributions of the indigenous peoples of the Americas: An anthropologist's suggestions for a multicultural curriculum.* Brunswick, Germany: International Textbook Research.

Kempf, S. (1997). *Finding solutions to hunger: Kids can make a difference: A sourcebook for middle and upper school teachers.* NECA. New York, NY.

Koch, P., Islas, A., Lee, H., Majumdar, D., Contento, I., Earl, C., ... Hoffman, E. (2010). Development of LiFESim: A social networking game to teach middle school students why and how to make healthful food and activity choices. *Journal of Nutrition Education and Behavior, 42*(4), s83.

Mace, D. (2010). Teaching about multicultural food to multicultural students in a multicultural school. *Geography,* 95(2), 80.

McNatt, Missy. (2009). Letters about the School Lunch Program. *Social Education, 73*(5), 198–202.

O'Connor, N. (2010, November 10) High school garden grows pride–and food; Windermere secondary's 'outdoor classroom' focuses on sustainability. *Vancouver Courier.* Retrieved from www.vancourier.com.

Slow Food USA. (2009). *Time for Lunch Lesson Plan.* Retrieved from http://www.slowfoodusa.org/downloads/campaigns/time_for_lunch-lessonplan.pdf

Wassermann, S. (2007). Big ideas: Let's have a famine! Connecting means and ends in teaching to big ideas. *Phi Delta Kappan, 89*(4), 290–297.

• CONTRIBUTORS •

Doug Davis is the Director of Food Service for the city of Burlington, Vermont, public schools, and he is a graduate of the Culinary Institute of America. He co-chairs the Food Service Directors Association of Vermont buying co-operative, is a consultant for the National Food Service Management Institute, and was a 10 year board member of the Vermont Campaign to End Childhood Hunger. Doug, who is frequently quoted in national media, has been working in child nutrition programs for over 18 years and has been actively involved in the farm-to-school movement for over 7 years. He resides on his farm in North Ferrisburg, Vermont.

Abraham P. DeLeon is an Assistant Professor at the University of Texas at San Antonio in the Department of Educational Leadership and Policy Studies. His critiques are grounded in critical and radical social theories and have been influenced by scholars such as Michel Foucault, Emma Goldman, Edward Said, Judith Butler, Jacques Derrida, and Henri Lefebvre. He has articles that appear in *The Social Studies*, *The Journal for Critical Education Policy Studies*, *Educational Studies*, *Theory and Research in Social Education*, *Critical Education*, and *Equity & Excellence in Education*. He also co-edited *Contemporary Anarchist Studies: An Introductory Anthology of Anarchy in the Academy* (Routledge, 2009).

Mi Ok Kang is a doctoral candidate in Curriculum and Instruction at the University of Wisconsin–Madison. Her current research focuses on how particular discourses and ideologies are represented and permeated through educational texts, restructuring and rescaling social relations in neo-liberal contexts. Drawing on a critical discourse analysis of educational policy texts in accordance with print media, she also examines the ways in which certain ideologies, cultures, and identities are legitimized in the field of education, supporting classed, raced, and gendered social structures. She recently was an editorial assistant on *The Routledge International Handbook of Critical Education* (2009) and *Global Crisis, Social Justice, and Education* (Routledge, 2009).

Irina Kovalskys, MD, is a specialist in pediatric nutrition. She is Assistant Professor at Favoloro University, Buenos Aires, Argentina. Her research concerns pediatric nutrition, overweight and obesity, epidemiology and public health. She is chair of the Committee of Obesity and Physical Activity, International Life Sciences Institute (ILSI), Argentina.

Catherine Lalonde is an Assistant Professor of Education at D'Youville College. She teaches "Critical Issues and Future Trends in Education" and "Multiculturalism and Cultural Diversity," she directs the D'Youville College Institutional Review Board (IRB), and her research interests include multicultural theory, social foundations of education, food production and consumption issues, and critical media literacy and pedagogy.

Marion Nestle is Paulette Goddard Professor in the Department of Nutrition, Food Studies, and Public Health at New York University. She is the author of *Food Politics: How the Food Industry Influences Nutrition and Health* (2nd ed., University of California, 2007), *What to Eat* (North Point, 2007), *Safe Food: Bacteria, Biotechnology, and Bioterrorism* (University of California, 2004), and *Pet Food Politics: The Chihuahua in the Coal Mine* (University of California, 2008).

Kristin Phillips is Assistant Professor in the Department of Teacher Education at Michigan State University. Her current research examines the challenges of local participation in national and international educational development agendas in rural East Africa. Her dissertation won the Gail P. Kelly dissertation award from the Comparative and International Education Society.

Sarah A. Robert is Assistant Professor at the University at Buffalo's Graduate School of Education. Her research and teaching explores the politics of education reform, particularly as it relates to teachers' work, social education, and gender equity. She has published related articles and book chapters in Argentina, Brazil, and the United States.

Jen Sandler is a Visiting Assistant Professor of Education at Bates College. Her research and teaching addresses the politics of knowledge, focusing specifically on activist efforts that teach people in power how to think about, understand, and address social suffering and inequality. She has published articles in *Routledge International Handbook of Critical Education* and *American Journal of Community Psychology*.

Marcus B. Weaver-Hightower is Associate Professor of Educational Foundations and Research at the University of North Dakota. He is the author of *The Politics of Policy in Boys' Education: Getting Boys "Right"* (Palgrave Macmillan, 2008) and co-editor of *The Problem with Boys' Education: Beyond the Backlash* (Routledge, 2009). His articles have appeared in *Educational Researcher, Review of Educational Research, Teachers College Record,* and *Discourse: Studies in the Cultural Politics of Education*, among others. His research interests include food politics, gender and education, educational policy, qualitative research methods, and the politics and sociology of education.

• INDEX •

0–9

4-H Organization, 76, 77, 85

A

Access to food, *see* Food deserts and food access
Activism, 3–4, 35, 43, 56–57, 105–106, 108, 188, 203–204
Advertising (marketing), 4, 5, 12, 14, 15, 50, 56, 58, 61, 66, 145, 147, 149, 153, 158, 188
After-school programs, 27-45, 147–159, 168
Allergies, *see* Food allergies
Anarchist theory, 108, 192–6
Animal rights, 1, 2, 14–15, 55–56, 183–197
Anthropomorphism, 185, 189–191
Argentina, 4, 94–116, 202
Australia, 46, 51, 59–68, 159

B

Black Panther Party, 28, 34–35, 43
Breakfast, *see* School breakfast
Brillat-Savarin, Jean, 6, 14, 113

C

Cafeterias, *see* School cafeterias
Cafeteria workers, *see* Food service workers
Child Nutrition Act (U.S.), 9, 10, 38, 56–59, 65
Childhood obesity, *see* Obesity
Class, *see* Socioeconomic status
Commodity program, 9, 53, 54, 57, 63, 144
Community Supported Agriculture (CSA), 167, 172–174
Contracted feeding services (private catering), 27, 29, 39–45, 49, 52, 59, 61, 124, 127–128, 134, 136
Cooking skills and training, 50, 61–62, 66, 76, 147–159, 181–182
Critical pedagogy, 148, 152, 185
Culture, 11,14, 34, 66, 109, 114, 155, 165, 167, 172, 183, 184, 190, 191
Curriculum
- animals in, 184, 187, 191, 201
- food and feeding as, 13–14, 25, 37, 42, 43, 204, 207, 209–212
- health, 115
- materials, 50
- school gardens, 71, 75–76, 79, 84–88, 179

D

Disciplinary practices, 41, 156–157, 185–191, 206
Discourses
- about nonhuman animals, 184, 188–190, 196
- about school cultivation, 72, 88-89
- about urban parents, 25–28, 32, 42
- about urban feeding, 29–30
- accountability, 187
- in policy ecologies, 6
- neoliberal, 62–63, 101
- of possibility, 194–195
- rights-based, 107, 114–116, 205
- social protection, 99
- South Korean school lunch, 121, 123, 125–130
- speciesism, 185
- vegetarian and vegan, 188

Dispositif, 189

E

Ecology of school food, 2, 6–11, 47–68, 95–116, 150, 155, 157, 158, 201–207
Educators, *see* Teachers and administrators
England (United Kingdom), 13, 15, 20, 46–54, 59, 62–68, 121, 150, 202
Environmental issues, 14–15, 76, 78, 86, 125, 166, 174, 180, 184, 188, 195

F

Farm-to-school programs, 76, 162–182
Fascism, 183–185, 192–197
Fast food, 11, 13–14, 36, 54, 145, 147, 158, 187, 188, 189, 192
Food
allergies, 12, 57, 155–156
deserts and food access, 76, 158, 166, 176
industry and corporations, 1, 3, 6, 9, 15, 36, 42, 50, 59-62, 63, 64, 66–68, 74, 97, 102, 109, 143–145, 149, 187, 188–189
preparation, 2, 5, 57, 61, 63, 66, 76, 102, 114, 147–159, 168. *See also* Cooking skills and training
purpose, *see* Purpose of feeding
restrictions and bans, 43, 51, 60, 63, 64–65, 150, 156
safety, 9, 10, 17, 98, 122, 127, 128, 175
security and insecurity, 16, 72–75, 77–78, 79, 94, 101, 109–110. *See also* Hunger
Food service workers, xiii, 2, 5, 6, 8, 11, 37, 39, 46, 49, 50–52, 64–66, 165–166, 168, 177–178, 193
Foucault, Michel, 183–196
Free for all meals (vs. free for selected), 120–136
Funding, 203, 204, 207
Argentina school feeding, 98, 99, 100, 101, 105–108, 113–114
breakfast, 35
for after-school programs, 27–30, 37
from vending and pouring rights, 13
research, 63, 67
reform groups, 65
School Food Trust, 49, 52–53
South Korea school feeding, 122–136
Fundraising, 15, 43, 59, 66, 162–163, 169–171,

G

Gardens, *see* School gardens and farms
General Agreement on Trade and Tariffs (GATT), 122, 125–126
Gender, 4–6, 14, 15, 27, 59, 63, 74–75, 84, 131, 145, 150, 153–155, 159, 190, 205
Globalization, 109, 126, 204

H

Healthy Kids Association (Australia), 46, 59–68
Healthy People 2010 (U.S.), 146
Healthy School Meals Act of 2010 (U.S.), 56–59
History of School Nutrition, 203
Argentina, 97–103
Australia, 59–62
Burlington, Vermont (U.S.), 167–169
England, 47–53
South Korea, 122–136
Tanzania, 79–87
United States, 33–42, 53–59
Hunger, xi, 1, 5, 13, 15–16, 27, 32, 35, 38–39, 40–41, 54, 59, 72–75, 77, 79, 84, 87, 90–91, 94–95, 102, 103–109, 113–114, 122, 165, 210–212. *See also* Food security and insecurity.
Hygiene, *see* Food safety

I

Ideology, 1, 11, 17, 39, 43–44, 62–63, 79, 130, 183, 184, 185–197, 204
Impacts of school food, 6, 12–16, 158
International Monetary Fund (IMF), 83

Intersectionality, 4, 14, 15, 155, 190–191, 194, 205
Interventions, 50, 56, 60–62, 74, 110, 111, 112, 147, 148–159

J

Junior Iron Chef competition, 162–165, 169, 177
"Junk" food, 1, 2, 11, 14, 42, 63, 102, 112, 201

L

Local food, xii, 7, 46, 114, 124, 125–126, 127, 134–136, 162–182
"Lunch ladies," *see* Food service workers
Lunchrooms, *see* School cafeterias

M

Malnutrition and undernourishment, xii, 12, 72, 73, 77, 78, 90, 94, 95, 102, 105, 107, 109, 110, 111, 113, 114, 115, 143
Manufacturers, *see* food industry,
Marketing, *see* advertising
Media, 5, 10, 51, 59, 60, 62, 63, 66, 100, 145, 149, 151, 158, 166, 181, 194
Milk and milk alternatives, 14, 48, 55, 57–58, 64, 98–99, 101, 102, 110, 122, 175, 188

N

National School Breakfast Program, *see* School breakfast
National School Lunch Program (NSLP; U.S.), xi, 9, 27, 33–36, 38, 40, 53–54, 57, 58, 172
Nutrition education, 13, 42, 43, 50, 56, 61, 75, 78, 97, 103, 110, 114, 115, 116, 147, 163, 201
Nutritional transition, 143–144, 109–110

O

Obesity and overweight, xii, 1, 10, 11, 12, 46, 56, 94, 96, 97, 109, 111–112, 113, 115, 143–144, 145, 147, 211
Oliver, Jamie, 10, 13, 26, 42, 49, 50, 55
Organic food, 1, 46, 120–136
Overproduction of food, 16, 144–145

P

Panopticon, 186
Parents, xii, 6, 8, 25–45, 50, 51, 59, 62, 79, 82, 86–87, 89, 104–106, 108, 125, 127, 136, 145, 154, 175, 201–203
Physicians Committee for Responsible Medicine (PCRM), 46, 55–59, 62–68
Plant-based diets, *see* Vegetarian and vegan diets
Policy ecologies, 6–7, 37, 97, 121, 196–197
Poverty, 5, 25, 26, 32, 33, 34, 39, 40, 44, 63, 74, 84, 90, 91, 95, 102, 104, 105, 115, 124, 128, 165, 207
Privatization of feeding, *see* Contracted feeding services
Purpose of feeding, 25, 28, 33, 34, 40–42, 44, 48, 53, 98–103, 202, 205–206

R

Race, 4, 9, 14, 15, 54, 63, 155, 190, 205
Reagan, Ronald (and administration), 35, 36, 54
Religion, 1, 4, 33, 63, 148, 150, 155–156
Resistance, 15, 64, 192–197

S

School
  breakfast, 15, 27, 34–36, 38–39, 40, 54, 77, 101, 168, 173–174, 201
  cafeterias, xiii, 9, 11, 13–14, 29, 48, 54, 63, 99-101, 108, 115,

129, 165, 166, 182, 184–193, 196, 204, 207
canteens, 15, 59-68, 112, 204
cultivation, *see* gardens and farms
food politics, 1, 2–6, 10–11, 27, 32–33, 42, 184, 2-3, 2-4, 207
gardens and farms, 17, 76, 77–91
lunch law (South Korea), 122–123, 128, 129, 134
School Food Trust (SFT), 13, 46, 49–53, 56, 60, 62–68, 202, 211
School-managed lunch systems, 124, 125, 127–128, 134
School Nutrition Association (SNA), 8, 9, 15, 55, 56, 66, 211
Selective social welfare, 131, 133
Social justice, 15–16, 19, 95, 148, 196
Socioeconomic status, 4–5, 14, 15, 54, 81, 120, 128, 131, 149, 150, 151, 155, 159, 190, 205
South Africa, 4–5, 62
South Korea, 15, 120–136, 206
Space, 42, 45, 115, 184–197, 203–204
Speciesism, 185, 189, 190
Subjectivity, 184, 187, 192, 194–195

T

Tanzania, 71–91, 202
Teachers and administrators, xii, 1, 8, 11–16, 28, 29, 33, 41, 50, 52, 63, 64, 67, 77, 128, 134, 146, 170, 196, 207
after-school programs, 29, 41
as activists, 96, 103–109, 125, 193, 194
bribes, 127
determining eligibility, 38, 129
professional development, 79, 168, 207
role in school gardens, 80-87
serving meals, 39
Thatcher, Margaret, 48
Transnational organizations, *see* International Monetary Fund, United Nations, World Bank, and World Trade Organization.

U

United Nations (UN), 3, 8, 9, 78, 84
United States Department of Agriculture (USDA), 8, 16, 18, 29, 36–37, 54, 55, 56, 57, 67, 76, 131, 144, 145, 211. *See also* Commodity program
Undernutrition, *see* Malnutrition and undernourishment
Universal social welfare, 131, 133

V

Vegetarian and vegan diets, 9, 55-57, 63, 64, 66, 67, 184–185, 188
Vending, 1, 13, 15, 26, 36, 64, 149, 165

W

World Bank, 8, 9, 83
World Trade Organization (WTO), 8, 10, 122